# INTERMEDIATE
# ORGANIC CHEMISTRY

# INTERMEDIATE ORGANIC CHEMISTRY

## Second Edition

**JOHN C. STOWELL**
Department of Chemistry
University of New Orleans
New Orleans, Louisiana

A Wiley-Interscience Publication
**JOHN WILEY & SONS, INC.**

New York  •  Chichester  •  Brisbane  •  Toronto  •  Singapore

*Library of Congress Cataloging in Publication Data:*
Stowell, John C. (John Charles), 1938–
    Intermediate organic chemistry / John C. Stowell.—2nd ed.
        p.    cm.
    "A Wiley-Interscience publication."
    Includes index.
    ISBN 0-471-57456-2 (acid-free)
    1. Chemistry, Organic.   I. Title.
QD251.2.S75   1993
547—dc20                                                    93-10028

Printed in the United States of America

10 9 8 7 6 5 4 3

# PREFACE

Consider the typical student who, having finished the two-semester introductory course in organic chemistry and then picking up an issue of the *Journal of Organic Chemistry*, finds the real world of the practicing chemist to be mostly out of reach, requiring a higher level of understanding. This text is intended to bridge that gap and equip a student to delve into new material.

There are two things to learn in studying organic chemistry. One is the actual chemistry, that is, the behavior of compounds of carbon in various circumstances. The other is the edifice of theory, vocabulary, and symbolism that has been erected to organize the facts. In first-year texts there is an emphasis on the latter with little connection to the actual observables. Thus students are able to answer subtle questions about reactions without knowing quite how the information is obtained. Chemistry is anchored in observations of specific cases, which can be obscured by the abstractions. For this reason this text includes specific cases with more details and literature references to illustrate the general principles. Understanding these cases is an exercise to ensure the understanding of those general principles in a concrete way.

Many of the problems begin with raw data, and require multistep thinking. Therefore the student must solve a problem from the beginning rather than from a half-finished setup, and more thought and puzzling is necessary.

This book is a selection. Out of the endless possibilities of reactions, a specific limited set was chosen and the chapters are consistent within that set. Likewise, the problems in the chapters were chosen for requiring only that set plus reactions from the introductory organic courses. These selections were made on the basis of their appearance in many current journal articles. Review references are given to aid a student who wishes to go further with these topics. Subjects that are generally well covered in introductory courses are omitted or briefly reviewed here. Advanced topics are treated to a functional level but not exhaustively.

For those familiar with the first edition, you will find that major changes have been made in all the chapters. New examples have been added, and many others have been replaced with better ones. The explanations have been elaborated and illuminated with more detail. The same 10 chapters have been retained, but new sections have been added, and material reorganized. Substantial updating has been done including stereochemical terminology, and the NMR chapter has been reframed in terms of pulsed high-field spectrometers. About twice as many exercise problems are available at the end of most chapters.

The author will provide copies of the *Instructor's Manual* to teachers using this book. Send a stamped (4 oz rate) self addressed 9 × 7 in. envelope with your request.

JOHN C. STOWELL

*New Orleans, Louisiana*
*October 1993*

# ACKNOWLEDGMENTS

The material on pages 26 and 27 from the *Ring System Handbook* is copyrighted by the American Chemical Society and is reprinted by permission. No further copying is allowed.

The material from *Beilstein's Handbuch der Organischen Chemie* on page 29 is copyrighted by Springer-Verlag and reprinted by permission.

Figure IX, Chapter 7 (an adaptation) is reprinted with permission from Nome, F.; Erbs, W.; Correia, V. R. *J. Org. Chem.* **1981**, *46*, 3802. Copyright 1981 American Chemical Society.

Figures XVIII–XX, Chapter 10 are adapted from Johnson, L. F. and Jankowski, W. C. *Carbon-13 NMR Spectra*, copyright© 1972 by John Wiley & Sons Inc.; reprinted by permission of John Wiley & Sons, Inc.

The graphs in Figures II and III of Chapter 8 were provided by S. L. Whittenburg, University of New Orleans.

The contour plots in Figure IV of Chapter 8 were provided by E. A. Boudreaux, University of New Orleans.

The plots in Figures IV, VI, and X of Chapter 7, and Figure XIV of Chapter 10 were provided by M. A. Polito, Tulane University.

The simulated spectra in Figures VI–VIII and XI of Chapter 10 were provided by R. F. Evilia, University of New Orleans.

Figure IX and Eq. 21 of Chapter 8, and Figures IX, X, XV, and XXI of Chapter 10 were provided by D. Lankin of Varian Associates, Park Ridge, Illinois.

Figure XVII of Chapter 10 was provided by B. Jursic, University of New Orleans.

I am grateful to Kenneth Andersen at the University of New Hampshire for reviewing the entire manuscript and providing many helpful suggestions.

J.C.S.

# CONTENTS

# LIST OF ABBREVIATIONS

| | |
|---|---|
| Ac | acetyl |
| AIBN | $(CH_3)_2C(CN)N=NC(CN)(CH_3)_2$ |
| Aq | aqueous |
| Bu | butyl |
| DME | 1,2-dimethoxyethane |
| DMF | dimethylformamide |
| DMSO | dimethylsulfoxide |
| Et | ethyl |
| HMPA | hexamethylphosphoric triamide |
| LDA | lithium diisopropylamide |
| Me | methyl |
| Ph | phenyl |
| Pr | propyl |
| rt | room temperature |
| THF | tetrahydrofuran |
| TMEDA | *N,N,N',N'*-tetramethylethylene diamine |
| Ts | *p*-toluenesulfonyl |

*The Journal of Organic Chemistry* uses many more abbreviations. They are defined in the *Guidelines for Authors* in the front of the first issue of each year.

# INTERMEDIATE
# ORGANIC CHEMISTRY

# 1

# READING NOMENCLATURE

Organic chemistry is understood in terms of molecular structures as represented pictorially. Cataloguing, writing, and speaking about these structures requires a nomenclature system, the basics of which you have already learned in your introductory course. To go further with the subject, you must begin reading journals, and this requires understanding of the nomenclature of complex molecules. This chapter presents a selection of compounds from recent issues to illustrate the translation of names to structural representations. The more difficult task of naming complex structures is not covered here because each person's needs will be specialized and can be found in nomenclature guides.[1-5] Most of the nomenclature rules are used to eliminate alternative names and arrive at a unique (or nearly so) name for a particular structure; thus, when beginning with names, you will need to know only a small selection of the rules in order to simply read the names and provide a structure. Although the subject of nomenclature is vast, these selections will enable you to understand many names in current journals.

## 1.1 ACYCLIC POLYFUNCTIONAL MOLECULES

**Methyl (3S,4S)-4-hydroxy-3-(phenylmethoxy)hex-5-enoate**

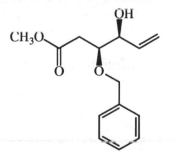

The space after methyl and the ''ate'' ending tells you this is a methyl ester. The acid is a six-carbon chain with a double bond between carbons 5 and 6. There is an alcohol function on carbon 4. There is a methoxy group on carbon 3 and a phenyl group on the carbon of the methoxy group. Carbons 3 and 4 are stereogenic atoms each with S configuration as designated. See Chapter 3 to review the meaning of wedges and dashed lines.

In all of these structures each vertex or end of a line (with no letter) represents a carbon atom with an appropriate number of hydrogens to complete tetravalency. A line that leads to a letter such as O is not a carbon–carbon bond, but a carbon–oxygen bond; that is, there are five carbons in ⌇⌇⌇OH.

Considering that free rotation exists about all the single bonds, and molecules can be viewed from any point, endless different representations of the same molecule can be drawn. One more is shown below.

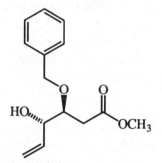

**3-(S)-*trans*-1-Iodo-1-octen-3-ol methoxyisopropyl ether**

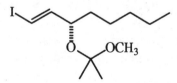

This is an example of a derivative name; that is, the first word is the complete name of an alcohol, and the other two words describe a derivatization where the alcohol is converted to an ether (ketal). In contrast, the previous compound, which is also a derivative of an alcohol, did not contain the full name of the alcohol.

**Ethyl (E,3R*,6R*)-3,6,8-trimethyl-8-[(trimethylsilyl)oxy]-7-oxo-4-nonenoate**

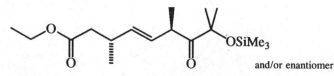

and/or enantiomer

This is the ethyl ester of a nine-carbon unsaturated acid with substituents. The *oxo* indicates that there is a keto function on carbon 7. Be careful to distinguish this from the prefix *oxa-*, which has a different meaning; see Section 1.6. The asterisks indicate that the configuration designation is not absolute but rather represents that stereoisomer and/or the enantiomer thereof. Thus this name represents the *R,R* and/or the *S,S* isomers, but not *R,S* or *S,R*. This designation excludes diastereomers, and is a common way to indicate a racemate.

## 1.2  CARBOCYCLIC COMPOUNDS (MONOCYCLIC)

**(1S,3R)-2,2-Dimethyl-3-(3-oxobutyl)cyclopropanecarboxylic acid**

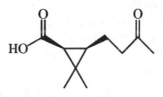

Let the ring atoms define the plane of the paper. The functional group that terminates the name, the carboxylic acid in this case, starts the ring atom numbering. The butyl substituent on the third carbon of the ring has a keto function on the third carbon of the butyl chain.

**[2S-(1E,2α,3α,5α)]−[3-(Acetyloxy)-2-hydroxy-2-methyl-5-methylethenyl)cyclohexylidene]acetic acid ethyl ester**

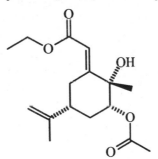

The *ylidene* indicates that the cyclohexyl is attached to the acetic acid by a double bond. That double-bonded ring atom is carbon 1, and the substituents on the ring are placed on the ring according to their locant numbers. The *E* indicates the geometry of the double bond. All the α substituents reside on one face of the ring, cis to each other. Any β substituents would reside on the opposite face of the ring, trans to the α substituents. Where two substituents are on the same ring atom, as on carbon 2 in this case, the Greek letter indicates the position of the higher-priority substituent. Here the hydroxy, acetyloxy, and methylethenyl are all cis to each other on the ring.

## 1.3  BRIDGED POLYCYCLIC STRUCTURES

**[1S-(2-*exo*,3-*endo*,7-*exo*)]-7-(1,1-Dimethylethyl)-3-nitro-2-phenylbicyclo[3.3.1]nonan-9-one**

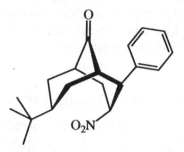

This molecule contains a ring with a bridge extending across it. It would require two bond breakings to open all rings (not counting the phenyl); thus it is termed *bicyclo*. Rather than viewing this as rings, two carbons are designated as bridgeheads (atoms 1 and 5 in this case) from which three bridges branch and recombine. These bridges contain three, three, and one carbons each, as indicated by the bracketed numbers separated by periods. All bicyclo compounds require three numbers in brackets, tricyclo require four, and so on. The numbering for locating substituents, heteroatoms, and unsaturation begins at one bridgehead and proceeds over the largest bridge to the other bridgehead and then continues to the next largest, and so on. The total number of atoms in the bridges and bridgeheads (excluding substituents) is given after the brackets, in this case as *non*. Finally, carbon 2 carries a phenyl that projects toward the smaller neighboring one-carbon bridge rather than the larger three-carbon bridge, as indicated by 2-*exo*. The 1,1-dimethylethyl group is also exo. This group is commonly called *tert-butyl*, but this is a *Chemical Abstracts* name built on linear groups. The use of prefixes *exo*, *endo*, *syn*, and *anti* to indicate such stereochemical choices is demonstrated generally as follows:

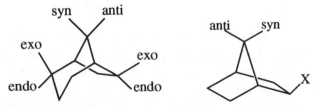

In tricyclic compounds, the relative stereochemistry among the four bridgeheads requires designation. Look at the largest possible ring in the molecule and consider the two faces of it. If there are no higher-priority substituents on the primary bridgehead atoms, the smallest bridge (but not a zero bridge) defines the $\alpha$ face. If the smallest bridge (not zero) at the secondary bridgeheads is on the same face of the large ring as the $\alpha$ defining one, it is also designated as $\alpha$; that is, the two are cis to each other.

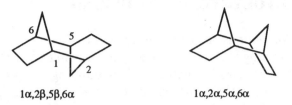

1$\alpha$,2$\beta$,5$\beta$,6$\alpha$                1$\alpha$,2$\alpha$,5$\alpha$,6$\alpha$

If they are trans, there will be two $\alpha$'s and two $\beta$'s as illustrated. If there is a zero bridge, the position of the bridgehead hydrogens is indicated with Greek letters as in the following example.

## (1$\alpha$,2$\beta$,5$\beta$,6$\alpha$)-Tricyclo[4.2.1.0$^{2,5}$]non-7-ene-3,4-dione

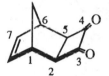

Starting with a pair of bridgeheads, draw the four-, two-, and one-carbon bridges. The zero bridge then connects carbons 2 and 5 as indicated by the superscripts, thus making them bridgeheads also. At bridgeheads 1 and 6 the smallest bridge is considered a substituent and given the $\alpha$ designation at both ends. At bridgeheads 2 and 5 the $\beta$'s indicate that the hydrogens are trans to the $\alpha$ bridge.

Sometimes a bridgehead substituent will have a higher priority than the smallest bridge thereon. The designation for that bridgehead will indicate the position, $\alpha$ or $\beta$, of that higher-priority substituent rather than the bridge as illustrated in the next example.

## (1$\alpha$,2$\beta$,4$\beta$,5$\beta$)-5-Hydroxytricyclo[3.3.2.0$^{2,4}$]deca-7,9-dien-6-one

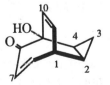

At bridgehead 1 the last numbered (smallest) bridge is considered a substituent on the largest ring and designated $\alpha$. The hydrogens at carbons 2 and 4 are trans to it and marked $\beta$. The OH group on carbon 5

is a higher-priority substituent than the C-9–C-10 bridge, and is trans to the bridge; thus it is labeled β.

## 1.4  FUSED POLYCYCLIC COMPOUNDS

Fused-ring compounds have a pair or pairs of adjacent carbon atoms common to two rings. Over 35 carbocyclic examples have trivial names, some of which need to be memorized as building blocks for names of more complex examples. The names end with *ene*, indicating a maximum number of alternating double bonds. A selection is illustrated, showing one resonance form for each.

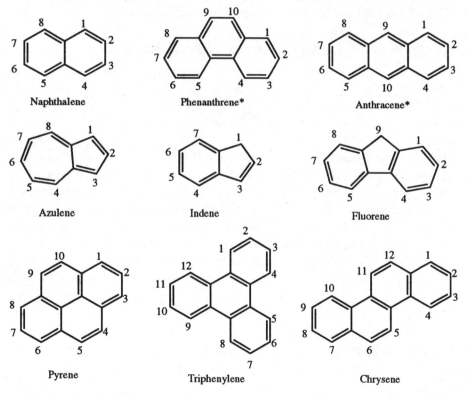

Naphthalene    Phenanthrene*    Anthracene*

Azulene    Indene    Fluorene

Pyrene    Triphenylene    Chrysene

* Exceptions to systematic numbering.

Fusing more rings onto one of these basic systems may give another one with a trivial name. If not, a name including the two rings or ring systems with bracketed locants is used, as in the following example.

## 7,12-Dimethylbenz[a]anthracene

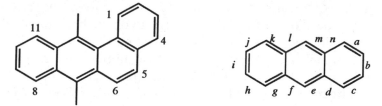

Since a side of the anthracene is shared, the sides are labeled *a*, *b*, *c*, and so on, where carbons numbered 1 and 2 constitute side *a* and 2 and 3 constitute side *b*, continuing in order for all sides. The earliest letter of the anthracene is used to indicate the side fused, and the ring fused to it appears first. The "o" ending of benzo is deleted here because it would be followed by a vowel.

The final combination is then renumbered to locate substituents, or sites of reduced unsaturation. To renumber, first orient the system so that a maximum number of fused rings are in a horizontal row. If there are two or more choices, place a maximum number of rings to the upper right. Then number clockwise starting with the carbon not involved in fusion in the most counterclockwise position in the uppermost or up-permost–farthest right ring. See the numbering in the systems with trivial names above. Atoms at the fusion sites, which could carry a substituent only if they were not $\pi$-bonded, are given the number of the previous position with an a, b, c, and so on. Where there is a choice, the numbers of the fusion carbons are minimized too; for example, 4b < 5a:

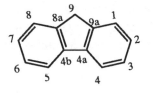

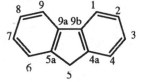

Correct                    Incorrect

## 1*H*-Benz[*cd*]azulene

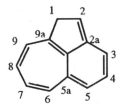

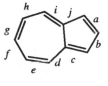

Azulene

First letter the sides of azulene. The benzo ring is fused to both the *c* and *d* faces as indicated in the brackets. Now reorient the system for numbering. The choice of which two rings go on the horizontal axis and which one in the upper position is determined by which orientation gives the smallest number to the first fusion atom; in this case 2a instead of 3a, or 4a. In molecules where one carbon remains without a double-bonding partner, it is denoted by *H*. This is called *indicated hydrogen* and is used even when an atom other than hydrogen is actually on that carbon in the molecule of interest.

### *trans*-1,2,3,4-Tetrahydrobenzo[*c*]phenanthrene-3,4-diol diacetate

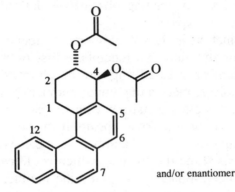

and/or enantiomer

The sides of phenanthrene are lettered; the carbons numbered 3 and 4 are the *c* side. As in several systems, phenanthrene is numbered in an exceptional way. A benzene ring is fused to the *c* side, and a new systematic numbering is made. Carbons 1–4 have hydrogens added to saturate two double bonds, and then carbons 3 and 4 have hydroxy groups substituted for hydrogens in a trans arrangement. Finally, it is the acetate ester at both alcohol sites.

### 6,13-Dihydro-5*H*-indeno[2,1-*a*]anthracene-7,12-diol

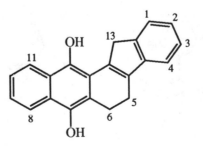

The bracketed *a* precedes anthracene therefore the sides of anthracene are lettered, while the 2,1 follows the indene therefore the sides of indene are numbered. All sides of benzo were the same but not so for indene; thus the bracket must include a designation for a side of it also. Furthermore, the direction of fusion must be shown since there are two choices for bringing the pair together fusing the same faces. The numbers in the bracket are carbons of indene fused in order 2, then 1. These are fused to the side of anthracene, with the number 1 carbon of anthracene constituting the number 2 carbon of indene; and the number 2 carbon of anthracene, the number 1 carbon of indene. The united system is then renumbered according to the rules and the substituents, added hydrogen, and indicated hydrogen placed accordingly. The indicated hydrogen is assigned the lowest-numbered atom not involved in double bonding.

The other direction of fusion is in 1*H*-indeno[1,2-*a*]anthracene:

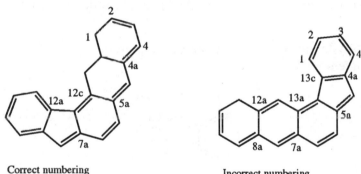

Correct numbering                    Incorrect numbering

Note that here a different orientation is used because it gives the lower fusion numbers:

Correct: 4a,5a,7a,8a,12a,12b,12c,13a
Incorrect: 4a,5a,7a,8a,12a,13a,13b,13c

## 1.5  SPIRO COMPOUNDS

In spiro compounds a single atom is common to two rings. There are two kinds of nomenclature for these. Where there are no fused rings present, the carbons of both rings are counted in one series, and the hydrocarbon name includes the carbons of both rings as in the following example.

### 4-Oxospiro[2.6]nonane-1-carboxylic acid, methyl ester

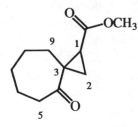

Of the nine carbons, one is the spiro atom, two join round to make a three-membered ring, and six finish a seven-membered ring, as indicated by the bracketed numbers. The locant numbers begin in the smaller ring at an atom adjacent to the spiro one, continue around the smaller ring including the spiro atom, and proceed around the larger ring.

When ring fusion is also present, the other nomenclature is used wherein the two ring systems that share the spiro atom are given in brackets with splicing locants as shown in the following example.

### 3′,4′-Dihydro-5′,8′-dimethylspiro[cyclopentane-1,2′(1′H)-naphthalene]

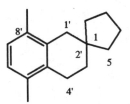

Two separate numbering systems are used. The unprimed number belongs to the ring nearest it in the brackets, the cyclopentane, while the primed numbers belong to their nearer neighbor in the brackets, naphthalene. The locant numbering in the fused system follows the usual pattern (Section 1.4) and is identified with primes. The -1,2′ indicates that the shared atom is number 1 of the cyclopentane ring and number 2′ of the naphthalene. The spiro linkage requires another naphthalene ring atom to be excluded from double bonding, in this case the 1′ as determined by the indicated hydrogen, 1′H.

## 1.6  HETEROCYCLIC COMPOUNDS (MONOCYCLIC)

Systematic and trivial names are both commonly in use for heterocyclic compounds. The systematic names consist of one or more prefixes from Table I with multipliers where needed designating the heteroatoms, fol-

**TABLE I.   Prefixes in Order of Decreasing Priority[a]**

| | |
|---|---|
| Oxygen | ox- |
| Sulfur | thi- |
| Selenium | selen- |
| Nitrogen | az- |
| Phosphorus | phosph- (or phosphor- before -in or -ine) |
| Silicon | sil- |
| Boron | bor- |

[a]An "a" is added after each prefix if followed by a consonant.

lowed by a suffix from Table II to give the ring size with an indication of the unsaturation. This is preceded by substituents. Thus, oxepin is a seven-membered ring including one oxygen and three double bonds. The ring size is explicit in some of the suffixes as *ep*, *oc*, *on*, and *ec*, which are derived from h*ep*tane, *oc*tane, n*on*ane, and d*ec*ane, respectively. Numbering begins with the element highest in Table I and continues in the direction that gives the lowest locant to the next heteroatom.

## 2-Propyl-2-(3-chloropropyl)-1,3-dioxolane

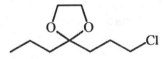

**TABLE II. Suffixes Indicating Ring Size**

| Atoms in the Ring | Containing Nitrogen | | Containing No Nitrogen | |
|---|---|---|---|---|
| | Maximally Unsaturated | Saturated | Maximally Unsaturated | Saturated |
| 3 | -irine | -iridine | -irine | -irane |
| 4 | -ete | -etidine | ete | -etane |
| 5 | -ole | -olidine | -ole | -olane |
| 6 | -ine | —[a] | -in | -ane |
| 7 | -epine | —[a] | -epin | -epane |
| 8 | -ocine | —[a] | -ocin | -ocane |
| 9 | -onine | —[a] | -onin | -onane |
| 10 | -ecine | —[a] | -ecin | -ecane |
| >10[b] | — | — | — | — |

[a]Use the unsaturated name preceded by "perhydro."
[b]Use the carbocyclic ring name with heteroatom replacement prefixes; oxa-, thia-, and so forth.

### 3,5-Di-*tert*-butyl-2-phenyl-1,4,2-dioxazolidine

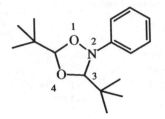

The numbering begins with oxygen for priority, choosing the one that gives the next heteroatom the lowest number. In the name the two oxygens come first and their numbers are first, also. The *-olidine* ending specifies a saturated five-membered ring.

### 4,7-Dihydro-2-methoxy-1-methyl-1*H*-azepine

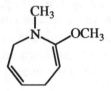

Azepine indicates an unsaturated seven-membered ring containing one nitrogen. One of the seven must have indicated *H*. The dihydro indicates that two others have additional hydrogen; therefore, there is one $\pi$ bond less than maximally unsaturated. Notice that the indicated hydrogen is assigned the lowest-numbered atom not in double bonding (the nitrogen) and *then* replaced by the substituent.

### (6*R*,14*R*)-6,14-Dimethyl-1,7-dioxa-4-(1-propylthio)cyclotetradec-11-yne-2,8-dione

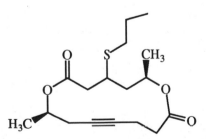

The ring is larger than 10 members; therefore, the hydrocarbon ring name cyclotetradecyne was used, modified by *1,7-dioxa*, which is a replacement of carbons 1 and 7 with oxygens. The *a* ending on *oxa* indicates replacement. The numbering begins at a heteroatom and proceeds to the other heteroatom by the shortest path. The stereochemistry at position 4 is unspecified.

Many five- and six-membered rings have trivial names that are preferred over the systematic. A selection of the more common ones is as follows:

**Unsaturated**

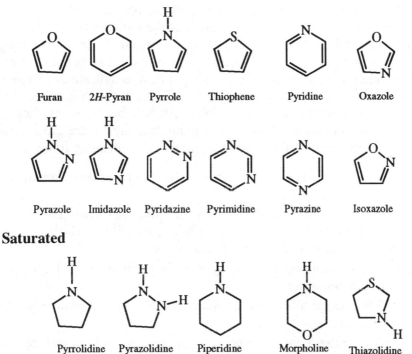

Furan      2*H*-Pyran      Pyrrole      Thiophene      Pyridine      Oxazole

Pyrazole      Imidazole      Pyridazine      Pyrimidine      Pyrazine      Isoxazole

**Saturated**

Pyrrolidine      Pyrazolidine      Piperidine      Morpholine      Thiazolidine

## 1.7    FUSED-RING HETEROCYCLIC COMPOUNDS

The names of the heterocycles in the previous section along with the rules for fusion in Section 1.4 are the basis for the following names.

### 3-Bromo-2-(2-chloroethenyl)benzo[*b*]thiophene

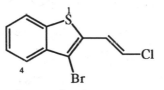

The final numbering begins at the most counterclockwise atom not involved in fusion, arranged to give the heteroatom the smallest possible number, in this case 1.

### 7-Bromo-1*H*-2,3-benzothiazine

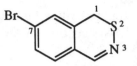

Sulfur is higher-priority than nitrogen, and the lowest number for it is 2. The 3 locates the nitrogen. In this case no bracketed site of fusion is specified because the fusion must precede the atom numbered 1. This is usual when there is more than one heteroatom and the fusion is simply benzo. The presence of one divalent atom in a six-membered ring excludes another atom from double bonding, thus the indicated hydrogen.

In choosing where to start numbering, and which direction to proceed around, a hierarchy of rules must be followed. Numbering always begins at a nonfused atom adjacent to a fused atom, but since there are several possible orientations for the molecule, a choice is made as follows:

1. Give the lowest possible number to the first heteroatom regardless of the priority of the atom.
2. If this allows two choices, choose the one that gives the second heteroatom the lowest number. If there are still two choices, minimize the number of the third, and so on.
3. If there are still two choices, give the lower number to the higher-priority heteroatom.
4. If all the above allow two choices, give the lowest number to the first fusion atom.
5. Finally, if there are still two choices, give the lowest numbers to the substituents.

Many benzo-fused heterocyclic compounds have trivial names for the combination of rings; for example:

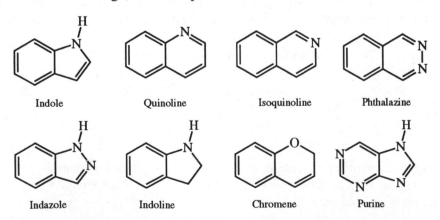

Indole        Quinoline        Isoquinoline        Phthalazine

Indazole        Indoline        Chromene        Purine

## 1,2-Dihydro-3-methylbenzo[ *f* ]quinoline

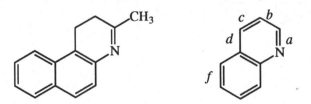

The sides of quinoline are lettered following the numbering system, and the benzo is fused to side *f*. The whole system is renumbered orienting as directed in Section 1.4, that is, maximum number of rings in a horizontal row, maximum in upper right, giving the heteroatom the lowest possible number, and numbering from the most counterclockwise nonfused atom in the upper right ring.

When two heterocyclic rings are fused, sides and direction of fusion are indicated in brackets as in the carbocyclic cases in Section 1.4. Examples of the six possible fusions between pyridine and furan are shown.

1. 6-Methylfuro[2,3-*b*]pyridine
2. 2,3-Dihydro-2,7-dimethylfuro[2,3-*c*]pyridine
3. 2,3-Dihydro-2-methylfuro[3,2-*b*]pyridine
4. 4-Bromofuro[3,2-*c*]pyridine
5. 5,7-Dihydrofuro[3,4-*b*]pyridine
6. Furo[3,4-*c*]pyridine-1,3-dione

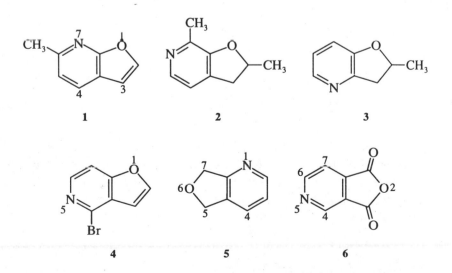

Furo[3,2-*b*] indicates that carbons 3 and 2 of the furan are the 2 and 3 carbons of the pyridine, respectively. The two-ring system is renumbered following the hierarchy of rules given above.

**Thieno[3,4-*c*]pyridine**

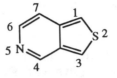

Thiophene as the first named in a fused system is shortened to thieno.

**2-Methyl-2*H*-thiazolo[4,5-*e*]-1,2-oxazine**

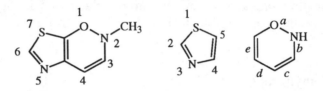

The name *thiazole*, like *oxazole*, means 1,3-thiazole. The atoms of thiazole are numbered and the sides of 1,2-oxazine are lettered. The fusion is drawn with atoms 4 and 5 of thiazole as atoms 5 and 6 of the oxazine. The system is then renumbered using the hierarchy of rules. You should flip your initial drawing about both the *x* and *y* axes to consider all four orientations to find the correct numbering.

**3*H*,5*H*-Thiazolo[3,4-*c*][1,3]oxazine**

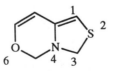

This is a different fusion of the same two rings. Note that this compound has brackets around the 1,3 of the oxazine while the previous structure did not have brackets around the 1,2. If the heteroatoms have the same numbers in the final structure as they had in the unfused oxazine, they are not bracketed, but if the final numbering gives new numbers to the heteroatoms, brackets are used. Notice that when a heteroatom is a fusion atom, it appears in the names of both rings, and is counted in the numbering.

## 1.8  BRIDGED AND SPIRO HETEROCYCLIC COMPOUNDS

Bridged and spiro heterocyclic compounds are named using the replacement nomenclature; that is, the hydrocarbon name is used with oxa, aza, and so on to substitute heteroatoms for carbons as was seen in large ring monocyclic compounds in Section 1.6.

**3-(2-Methylphenyl)-3,8-diazabicyclo[3.2.1]octane**

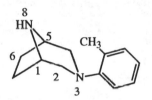

**2-(2′,6′-Dimethoxyphenyl)-1,3-oxazaspiro[5.5]undeca-2,7,10-trien-9-one**

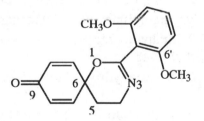

Many of the names that you will see in journals should be understandable by analogy from the examples studied here, but certainly there are many more complicated cases beyond the scope of this chapter. The chapter references should be consulted for them.

## PROBLEMS

Draw complete structures for each of the following compounds.

1. 2,4-Dimethylbenzo[ g]quinoline
2. 2-(Bromomethyl)-4,7-dimethoxyfuro[2,3-d ]pyridazine

3. Spiro[cyclopentane-1,3′-bicyclo[4.1.0]heptane]

4. (1R,3R,5S)-endo-1,3-Dimethyl-2,9-dioxabicyclo[3.3.1]nonane

5. 6-Methoxyspiro[4,5]decane

6. 1-Benzoyl-2-phenylaziridine

7. 3-Butyryl-2-(3-chloropropyl)-1-(methoxycarbonyl)-1,2-dihydropyridine

8. 7-Methyl-7H-benzo[c]fluorene-7-carboxylic acid

9. endo-8-Hydroxybicyclo[5.3.1]undecan-11-one

10. (1α,2α,5α,6β)-Tricyclo[4.3.1.1$^{2,5}$]undecane-11-one

11. (1α,2β,3α,5α)-6,6-Dimethylbicyclo[3.1.1]heptane-2,3-diol

12. [1R-(1α,2β,6α)]-4,7,7-Trimethylbicyclo[4.1.0]hept-3-en-2-ol

13. 6-(Benzyloxycarbonyl)-3-cyano-4-chloro-6-azabicyclo[3.2.1]oct-3-ene

14. Thieno[3,4-b]pyridine

15. 2H-3,1-Benzothiazine

16. Pyrrolo[2,1-c][1,2,4]triazine

17. 5′-Acetyl-4′-amino-1,3-dihydro-6′-methyl-1,3-dioxospiro[2H-indene-2,2′-2H-pyran]

18. 5-Methylbenzo[b]chrysene

19. (1α,2β,5β,6α)-Tricyclo[4.2.1.1$^{2,5}$]deca-3,7-diene-9,10-dione

20. (exo,syn)-2-(1-Pyrrolidinyl)bicyclo[3.3.1]nonan-9-ol

21. 3-Amino-5,6,8,9-tetrahydro-7H-pyrazino[2,3-d]azepine-2,7-dicarboxylic acid diethyl ester

22. 2-Benzoyl-1,6,7,11b-tetrahydro-2H-pyrazino[2,1-a]isoquinoline-3,4-dione

23. (1α,4α,5α,6α) - 3,4 - dichloro - 6 - (2,2 - dichlorocyclopropyl)bicy - clo[3.2.1]-oct-2-ene

24. 8-bromo-7-chloro-2-(ethylthio)-4H,5H-pyrano[3,4-e]-1,3-oxazine-4,5-dione

25. [1R-(1α,2β,4α)]-4-chloro-2-methylcyclohexanecarboxylic acid

26. 1′,2′-dihydro-4-methyl-2′-oxospiro[4-cyclohexene-1,3′-[3H]indole]-2,2-dicarboxylic acid diethyl ester

## REFERENCES

1. Rigaudy, J.; Klesney, S. P. (preparers). *Nomenclature of Organic Chemistry*, International Union of Pure and Applied Chemistry, Pergamon Press, Oxford, 1979.

2. *Index Guide, Chemical Abstracts*, American Chemical Society, Columbus, OH, 1987–1991, Appendix IV.

3. McNaught, A. D. The Nomenclature of Heterocycles, *Adv. Heterocyclic Chem.* **1976,** 20, 176–319.

4. Cahn, R. S.; Dermer, O. C. *Introduction to Chemical Nomenclature*, 5th ed., Butterworths, London, 1979.

5. Godly, E. W. *Naming Organic Compounds: A Systematic Instruction Manual*, Wiley, New York, 1989.

# 2

# SEARCHING THE LITERATURE

A truly vast and ever-growing body of organic chemical information is recorded in the chemical literature.[1-5] In a research library practically all of it can be searched surprisingly quickly with the use of *Chemical Abstracts*, *Beilsteins Handbook of Organic Chemistry*, and the review literature.

*Chemical Abstracts* (CA) covers over 12,000 periodicals, patents, and other sources and produces brief summaries of the information in each article with a bibliographic heading. These appear in weekly issues. Six months of these issues constitute a volume that is accompanied by six indexes: *General Subject Index*, *Chemical Substance Index*, *Formula Index*, *Index of Ring Systems*, *Author Index*, and *Patent Index*.[6] January to June of 1993 is volume 118. Ten volumes (5 years' coverage) are combined in the *Collective Indexes*; the eleventh collective volumes are 96–105 (1982–1986). The eleventh *Collective Index* comprises 93 bound books and occupies 17 ft of shelf space. The indexes are based on the entire original document and will include compounds that are not specifically mentioned in the abstract. The twelfth *Collective Index* will be available on CD-ROM as well as the usual printed form, although the disks will not include the structure diagrams.

For most purposes it is best to begin searching in the latest available volume indexes and continue to earlier volumes and into the *Collective Indexes*. Often the earlier literature on a subject or a substance is referenced or summarized in the more recent papers, and it will simplify your search. Before the ninth *Collective Indexes*, there were decennial indexes back to volume 1, 1907. In the following sections the use of these indexes are illustrated with examples. It would be best to go to the library and follow along with these examples in the various volumes.

## 2.1 GENERAL SUBJECT SEARCH

The procedure for searching by subject is illustrated here by answering the question, "Is the structure of the mating pheromone of the Japanese beetle *Popillia japonica* known, and if so, has it been made synthetically?" The *General Subject Index* entries are made up of a controlled vocabulary to prevent scattering of references under several entries. Of the several words in this question, only certain ones are of value in the index. The first step is to examine the *Index Guide*, which lists the general subject index headings and chemical substance names used in the indexes, and refers the many alternative words and nonsystematic chemical names to the names used in the indexes. There is an *Index Guide* corresponding to each collective period. The latest covers the twelfth *Collective Index* period (1987–1991) in two parts. A 1992 *Index Guide* is available for the 1992–1994 period, which will be revised three times culminating in 1997 for the thirteenth *Collective Indexes*. In the 1992 *Index Guide* under "Pheromones" we find that studies of these as a class and of new pheromones are given under this term but studies of known ones are at their specific headings. Under *Popillia japonica* we find "see Japanese beetle." Turning now to the latest *General Subject Index* volume 116, under "Japanese beetle" we find a reference concerning inhibition of feeding but nothing about a pheromone. Continuing back, we find a reference to a patent **115:** P278160b on the preparation of the pheromone. Continuing further, we find references on control, attractants, insecticides, and finally "pheromone of, asym. synthesis of, **100:** 102990y." The number refers to an abstract; consulting that abstract we find the structure pictured, the name $(R,Z)$-$(-)$-5-(1-decenyl)oxacyclopentan-2-one, a reference to the journal *Agric. Biol. Chem.*, the fact that it was synthesized, and a key step in that process. The original document should next be consulted for details and references to earlier syntheses. A few more volumes should be searched to catch any recent papers missed by those authors.

Some abstract numbers in the indexes are preceded by a letter "B" indicating a book, "P" a patent, or "R" a review.

You should prepare a checklist record of all your literature searches, even if you find nothing on the subject, in order to know how far you went with a search and to save having to repeat later.

If you need to continue into earlier literature with your subject search, check the earlier *Index Guides* in case changes in vocabulary were made. In the eighth *Collective Index* and earlier, the *Subject Index* contains both chemical substances and subjects alphabetized together. (Earlier than the eighth *Collective Index*, there are no index guides. Appropriate "see" and "see also" references are given in the subject index itself.)

In 1966 and earlier the indexes give column numbers instead of abstract numbers. For example, **65:** 18621e refers to a column in the abstracts and the letter "e" refers to a distance down the column on the page.

For a more general subject such as the synthesis of all insect pheromones it may be more profitable to start with a review article. These may be quickly located by using *Index of Reviews in Organic Chemistry* published by the Royal Society of Chemistry (London). This consists of annual and collective issues covering material back to the early 1960s. The recent issues have two sections: Section 1 covers compounds, and Section 2 covers processes and phenomena. The reviews from 1940 to 1964 are found in three volumes by Kharasch and Wolf, *Index to Reviews, Symposia*, and *Monographs in Organic Chemistry*. A number of reviews are found under insect in Section 2 and under pheromones in Section 1. The reviews are also found in *Chemical Abstracts* where they are recognized by a capital "R" preceding the abstract number; for example, **95:** R79987v leads to a 183-page review with 486 references on the synthesis of insect pheromones. *J. Org. Chem.* includes an indexed list of recent reviews four times each year.

## 2.2 CHEMICAL SUBSTANCE NAME SEARCH

There are many possible names, common and systematic, that may be used for most compounds. Each compound appears at only one place in a volume of the *Chemical Substance Index*; therefore, you must find the CA name. That name will consist of a parent followed by modifiers, each listed alphabetically. Suppose you were interested in 1-benzyl-2-naphthoic acid. The parent name would be naphthoic acid. Check the *Index Guide* to see what parent name CA uses. In the 1992 issue we find under "Naphthoic acid:" "See Naphthalenecarboxylic acid [*1320-04-3*]." In the eleventh collective *Chemical Substance Index* we find 1-and 2-naphthalenecarboxylic acid, each with long lists of modifiers. In the list of modifiers we find "――, 1-(phenylmethyl)-[*73194-80-6*], **98:** 197730r," and in the tenth collective it appears again: **92:** 215181v; and the next entry is "prepn. and cyclocondensation of, **95:**

219898u.'' The abstracts themselves should then be consulted and from these the original papers may be found. Other volumes of the index should also be checked starting with the most recent. If in proceeding to earlier literature you lose track and suspect a nomenclature change, consult the *Index Guide* of that period or the *Formula Index*.

Clues to the modifier names can be found in the *Index Guide* also; for example, under ''Benzyl hydroperoxide'' we find ''See hydroperoxide, phenylmethyl.'' *tert*-Butyl is also not now used by CA, but an idea can be found, for example, under ''*tert*-Butyl hydroperoxide'' we find ''See hydroperoxide, 1,1-dimethylethyl.''

Since the *Chemical Substance Index* is alphabetized by the parent name first, it can be used to find examples of a class of compounds where you are unsure as to what cases might be known. For instance, if you were looking for an example of the structure

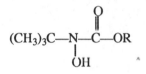

where the R group could be any size, the formula index would require selection of a particular case that might not happen to be known. However, under ''carbamic acid (1,1-dimethylethyl)hydroxy-'' we find in the tenth collective *Chemical Substance Index* ethyl ester, methyl ester, and phenyl ester.

The *Index Guide* is also useful for finding the true identity of commercial materials such as the pharmaceutical compound bufexamac. The *Guide* gives a complete name: ''Benzeneacetamide, 4-butoxy-*N*-hydroxy [*2438-72-4*].'' The numbers in brackets are *registry numbers* which are useful in computer searching; see Section 2.8.

If you plan to search for a rare compound of fair complexity, it may be difficult for an infrequent user to devise the CA systematic name in order to use the *Chemical Substance Index*. In these cases it may be preferable to start with the *Formula Index*.

## 2.3  MOLECULAR FORMULA SEARCH

The elemental composition of each substance is specified in the order carbon, hydrogen, and other elements in alphabetic order. These formulas are arranged in order of increasing numbers of carbons and for a given number of carbons, in order of increasing numbers of hydrogens, and so on for the other elements. If the compound is a salt such as an amine hydrochloride, acetate, or sulfate, it will be found under the formula for the free amine. Quaternary ammonium salts are under the formula for the cation, omitting the anion.

For example, let's try to find the preparation and properties of the enol acetate of chloroacetone:

$$\underset{\displaystyle ClCH_2C=CH_2}{\overset{\displaystyle \overset{O}{\overset{\|}{OCCH_3}}}{\big|}}$$

The molecular formula is $C_5H_7ClO_2$. Turning to the latest *Formula Index*, volume 117 shows 8 substances with this formula. Simply look at the names and quickly reject almost all of them; for example, 2-propenoic acid, modified, is not relevant. Careful examination of the propenols shows that this compound is not listed here. Proceeding to earlier volumes, we eventually locate one reference as tabulated here:

| | |
|---|---|
| *117* | No entry |
| *116* | No entry |
| *115* | No entry |
| *114* | No entry |
| *113* | No entry |
| *112* | No entry |
| *111* | No entry |
| *110* | No entry |
| *109* | No entry |
| *108* | No entry |
| *107* | No entry |
| *106* | No entry |
| 11th coll. (*96–105*) | No entry |
| 10th coll. (*86–95*) | No entry |
| 9th coll. | No entry |
| 8th coll. | No entry |
| 7th coll. | 1-Propen-2-ol, 3-chloro, acetate **56:** 7123d |
| 6th coll. | No entry |
| *41–50* | No entry |
| *14–40* | No entry |

The CA *Formula Index* does not extend earlier than 1920, but *Beilstein* has a formula index that does; see Section 2.6.

Consulting the abstract to the only entry found, we find the reference is Euranto, E.; Kujanpaa, T. *Acta Chem. Scand.* **1961**, *15*, 1209. The article gives no references to other papers for this compound, but it includes the preparation and physical properties of 3-chloropropen-(1)-yl-(2) acetate.

The *Formula Index* gives only the abstract and registry numbers (and indicates patents with a *P*) but no indication of the paper content. For substances where many references exist, it may be better to use the *Formula Index* to get the CA name and then to go to the *Chemical Substance Index* where modifiers are given, to select which abstracts you want to read.

In searching the *Formula Index* it is important to watch for nomenclature changes. For example, in the tenth collective *Formula Index* under $C_{10}H_{18}O_2$ we find the following structure named 2(3*H*)-furanone, dihydro-3-methyl-5-pentyl, **95:** P115267w:

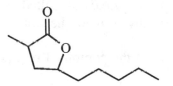

In the abstract itself it is referred to as an $\alpha$-methyl-$\gamma$-alkyl-$\gamma$-butyro-lactone. In the seventh collective *Formula Index* the name is "nonanoic acid, 4-hydroxy-2-methyl-$\gamma$-lactone **59:** 11205d." The change in the locant for the methyl group should not be overlooked. Once again, you needn't know all these nomenclature changes but when the furanones disappear suddenly at the seventh *Collective Index*, it is time to look at the whole list under the formula and recognize the older nomenclature.

## 2.4  RING SYSTEM SEARCH

Sometimes the formula that you are searching leads to names of many complex cyclic compounds and combing through them for the one you want is tedious. You can locate the particular parent name for the ring system first and then search in the *Formula* or *Chemical Substance Index* with it. This is done by using the *Ring Systems Handbook*. At this writing, the latest edition is dated 1988. Supplements are issued twice a year, and the latest is *Cumulative Supplement 8*, dated November 1992, which is to be used together with the 1988 edition. The next edition of the *Handbook* will appear in 1993.

There may be occasions when you want to find an example of a certain ring system regardless of what substituents may be present. Here, too, the *Handbook* should be used. With the *Ring Systems Handbook* you can find the parent name, and then search for it in the *Chemical Substance Index*, where all the substituted cases will be gathered under that heading regardless of their formulas. Again it is advisable to begin with the latest semiannual volume of the *Index* and continue back. The *Ring System Handbook* includes monocyclic and fused, bridged, and spiro polycyclic compounds.

To introduce yourself to this *Handbook*, you should go to the library and find an example compound containing the following ring system.

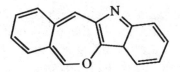

Begin with that part of the *Handbook* called the *Ring Formula Index*, with the following hierarchy of index headings and subheadings.

1. Elemental formula of the combined ring atoms: $C_{16}NO$
2. Number of rings: 4
3. Elemental formula of each ring starting with the smallest:

   Ring size         5,   6,   6,    7
   Ring formula   $C_4N-C_6-C_6-C_6O$

These individual ring formulas are columned in the usual formula index order.

In the 1988 edition we find
   $C_{16}NO$: 2 RINGS
      Many entries
   $C_{16}NO$: 3 RINGS
      many entries
   $C_{16}NO$: 4 RINGS
      $C_3NO-C_4N-C_6-C_7$
         many entries
      $C_4N-C_6-C_6-C_6O$
         1*H*-[1]Benzoxepino[5,4-*b*]indole [*RF 44948*]

1*H*-[2]Benzoxepino[4,3-*f*]indole [*RF 44949*]

1*H*-[2]Benzoxepino[4,3-*b*]indole [*RF 44950*]

1*H*-[2]Benzoxepino[4,3-*e*]indole [*RF 44951*]

1*H*-Dibenz[2,3:6,7]oxepino[4,5-*c*]pyrrole [*RF 44953*]

Oxepino[3,2-*d*]carbazole [*RF 44954*]

$C_4NO-C_5N-C_6-C_6$*

We may recognize the name of the one sought or limit it to a few which we then view structurally in *Ring Systems File II* filed by the *RF* numbers. In this case all six compounds under $C_4N-C_6-C_6-C_6O$ are found together as numbers *RF 44948–RF 44954*. We easily recognize the target ring system as *RF 44950*. The name found there is now searched in the *Chemical Substance Index*. An example CA citation is given also in the *File*, here **88:** 6775v.

## 2.5  AUTHOR NAME SEARCH

Occasionally you will want to see what was done next in a certain research group. This may be found using the CA *Author Index*. All the authors on each paper are indexed. The second, third, and so on will refer back to the first author, where you will find a brief indication of the content of each paper and an abstract number. The alphabetization is not the same as in a telephone directory because many papers give only the last name (surname) and initials. The order is alphabetized by last name and then first initial, and then second initial. An author index accompanies each CA volume and collective index. (CA does maintain a name authority file and will use the full name of the author if it is known to them. Thus in the author index; you may find the author entered as Smith, James W., even though it is given on the paper as J. Smith.)

You may find what other workers have since done with information from a particular paper by using the *Science Citation Index* published by the Institute for Scientific Information, Philadelphia, and covering the years 1961 and 1964 to present. The annual issues list all the references made to papers in a one-year period. Last names of the first

*Ring Systems Handbook, Ring Formula Index*, Copyright © 1988 by the American Chemical Society.

authors of all papers referred to are alphabetized, and the papers by each author referred to are listed chronologically. After each paper is given a list of new papers that cite that original paper. The current year is covered by temporary 2-month issues, which are later replaced with the complete year index.

## 2.6  BEILSTEINS HANDBOOK OF ORGANIC CHEMISTRY

If you are interested in the preparation and/or properties of a particular organic compound, searching *Chemical Abstracts* may be frustrating because many of the references you uncover will give a use for the compound and not the data you are seeking. This particular information may be found quickly in *Beilsteins Handbuch der Organischen Chemie*. *Beilstein* is an organized collection of preparations and properties of organic compounds that were known before 1960. The fourth edition (*Vierte Auflage*) consists of a basic series (*Hauptwerk*) covering work up to 1909, and four supplementary series (*Erganzungswerk*) covering the literature to 1959. A fifth supplement will eventually cover 1960–1979.

The *Handbook* consists of 29 volumes. The main divisions are as follows: acyclic compounds, volumes 1–4; carbocyclic compounds, volumes 5–16; heterocyclic compounds, volumes 17–27; *General Subject Index*, volume 28; and *General Formula Index*, volume 29. The *General Formula Index* is complete in the second supplement and covers the *Hauptwerk* and first and second *Erganzungswerk*. Many of these volumes consist of sets of bound subvolumes. It is hardly a "handbook" now since it occupies over 45 ft of shelf space.

The compounds are arranged in the volumes according to the rules of the *Beilstein* system, which allow you to search directly in the volumes without using indexes, finding similar compounds located together. These rules are beyond the scope of this chapter but are available elsewhere.[7] The example search shown below begins instead with a formula index.

The illustrative search example is as follows. Find the melting point and preparations of *N,N'*-diisopropylurea, $C_7H_{16}N_2O$. Consulting the *General-Formelregister*, volume 29, we find 11 isomeric ureas listed, including "*N,N'*-Diisopropyl-harnstoff **4**, 155, II 631." This indicates that it is in volume 4 in the *Hauptwerk* on page 155, not covered in the first *Erganzungswerk*, but in the second *Erganzungswerk* on page 631. Consulting these pages, we find ameisensäuremethylester und Isopropylamin in Wasser (THOMAS, *R.* **9**, 71). — Flüssig. Kp: 165,5°. D$^{15}$: 0,981.

**N-Isopropyl-harnstoff** C$_4$H$_{10}$ON$_2$ = (CH$_3$)$_2$CH · NH · CO · NH$_2$. *B.* Durch Reduktion der Verbindung (CH$_3$)$_2$C——N · CO · NH$_2$ (Syst. No. 4190) mit Aluminiumamalgam (CONDUCHÉ, *A.* O $^´$
*ch.* [8] **13**, 65). — Nadeln. F: 154°. Löslich in Wasser, Alkohol, Chloroform, siedendem Benzol und Aceton, weniger in Essigester, schwer in kaltem Äther und kaltem Benzol.

**N.N'-Diisopropyl-harnstoff, symm.** Diisopropylharnstoff C$_7$H$_{16}$ON$_2$ = (CH$_3$)$_2$CH · NH · CO · NH · CH(CH$_3$)$_2$. *B.* Aus N-Brom-isobutyramid durch Erhitzen mit Na$_2$CO$_3$, neben Isopropylisocyanat (A. W. HOFMANN, *B.* **15**, 756). — Nadeln (aus Alkohol). F: 192°. Unlöslich in Wasser und Äther.

Reprinted with permission from Prager, B.; Jacobson, P. *Hauptwerk*, Vol. 4 *Beilsteins Handbuch der Organischen Chemie.* Copyright 1922 Springer-Verlag, Heidelberg.

Brom auf Malonsäure-bis-isopropylamid in heißem Eisessig (WEST, *Soc.* **127**. 751). — Nadeln (aus Alkohol). F: 204°. — Geschwindigkeit der Reaktion mit Jodwasserstoffsäure in 4% Wasser und 2% Essigsäure enthaltendem Methanol bei 25° und 30,2°: W.

**N.N'-Diisopropyl-harnstoff** C$_7$H$_{16}$ON$_2$ = (CH$_3$)$_2$CH · NH · CO · NH · CH(CH$_3$)$_2$ (H 155). *B.* Neben Spuren von N-Isopropyl-N'-isobutyryl-harnstoff bei der Einw. von 1 Mol 10%iger Natronlauge auf N-Chlor-isobutyramid ohne Kühlung (ROBERTS, *Soc.* **123**, 2782). Beim Erwärmen einer wäßr. Lösung des Kalium- oder Natriumsalzes des Isobutyrhydroxamsäurebenzoats (JONES, SCOTT, *Am. Soc.* **44**, 421). — Krystalle (aus verd. Alkohol).

Reprinted with permission from Richter, F. *Zweites Erganzungswerk*, Vol. 4, *Beilsteins Handbuch der Organischen Chemie.* Copyright 1942 Springer-Verlag, Heidelberg.

We now locate it in E III and/or E IV, which are not included in that index. Knowing that it is in volume 4 from the first formula index, we now turn to the new *Gesamtregister*, volume 4 formula index, which covers all parts of volume 4, including the fourth supplement. Under the formula we find "Harnstoff, *N,N'*-Diisopropyl-, **4** 155b, II 631a, III 277a, IV 521." The letters indicate how far down the page to look; the first compound on the page is a, the second b, the eighth h, and so on. The melting point is 197°C, and quite a variety of syntheses are given with references.

In those volumes where the new formula index is not yet available, you can locate compounds in E II and E IV by examining the heading on the right-hand pages in these supplements, looking for those pages that correspond to ones in the earlier issues. For example, in volume 13, on top of page 141 we find

*Reprinted by permission from Luckenbach R. *Viertes Erganzungswerk*, Vol. 13, *Beilsteins Handbuch der Organischen Chemie.* Copyright 1985, Springer-Verlag, Heidelberg.

Material on this page is an extension of that on pages 96–97 of the *Hauptwerk* and pages 169–172 of the third supplement of this volume.

There is also a formula and subject index in each individual subvolume in the third and fourth supplements.

Parts of the *Fifth Supplementary Series* have appeared, covering the literature from 1960 through 1979. This series is in English, while the earlier ones are all in German.

If the compound you are seeking does not appear before 1929, you will need to use the *Beilstein* system to determine in which volume it should appear in the third and later supplements. For a concise introduction, see the booklet *How to Use Beilstein*, Beilstein Institute, Springer-Verlag, Berlin, 1978.

## 2.7   GENERAL SOURCES

Thus far we have discussed finding original references to chemical information. A general idea of the reactions and the properties of classes of compounds is more readily obtained from compilations of organic chemistry. These are encyclopedic works spanning the entire field. One of the best is *Methoden der Organischen Chemie*, known as "Houben-Weyl" for the editors of the first edition. The fourth edition was completed in 1985 with 65 volumes and a general index. Since then, a series of supplementary volumes (Erweiterungs- und Folgebänden) have issued. This is an organized, completely referenced, very detailed collection of methods of preparing essentially all classes of organic compounds plus their reactions. It includes selected experimental details and extensive tables of examples. The publisher is Georg Thieme Verlag, Stuttgart.

*Comprehensive Organic Chemistry* is a five-volume work plus a sixth volume of indexes again spanning the whole field and providing many leading references. The editorial board was chaired by D. Barton and W. D. Ollis. It was published complete and contemporaneous in 1979 by Pergamon Press, Oxford.

There are several other such compendia and also hundreds of individual monographs available on specialized topics within the field.

## 2.8   ONLINE COMPUTER SEARCHING

Many of the preceding sources may be searched online through a computer logged on to one of the several suppliers of information including BRS Information Technologies; DIALOG Information Systems, Inc.;

ORBIT Search Service; and STN International (jointly operated by FIZ Karlsruhe, The Japan Information Center of Science and Technology, and Chemical Abstracts Service). These vendors carry many databases covering other fields besides chemistry. The user pays a search fee through an account with the vendor plus a telecommunications fee. The fee is based on connect time, search terms, and answer displays.

STN carries over 150 databases called *files*, including four based in printed *Chemical Abstracts*.

1. *Registry File.* This file contains over 12 million substances from the sixth *Collective Index* to the present, as well as a limited number from earlier indexing periods. The file is updated each week with 8000–14,000 new substances. To search for a particular compound, or an example of a structural class, you should begin in the Registry File. You may enter a drawn structural formula, molecular formula, substance name, partial structure, name fragment, or a CAS Registry Number.® The answer for each compound will include the CAS Registry Number, a structural formula, substance name, synonyms, molecular formula, the 10 most recent references in the CA File, and whether there are any references in CAOLD and CA Previews. You can opt for the abstract numbers, which you can then examine in the printed abstracts. You may further opt for the bibliographic data and go directly to the original documents. If there are more than 10, you need to transfer to the CA file.

2. *CA File.* This file contains the complete bibliographical information for over 10.6 million documents from all the abstracts in CA since 1967, plus all the abstract texts since 1970. Abstracts for earlier periods are being added regularly. The CA file is updated every 2 weeks with 14,000–18,000 records. You may enter a CAS Registry Number, subject term, or item of bibliographical information. This file contains not only the controlled vocabulary but also the document titles, the keywords from the individual CA issues, and the subheading phrases (text modifiers) from the indexes, all of which can be searched. Even the text of the abstract can now be searched. The answer will be a list of CA abstract numbers, and if selected, bibliographic information, and the abstract text. It is particularly valuable to search with a combination of two or more Registry Numbers and/or subject terms. This allows the use of broad terms to find small sets of answers.

3. *CAOLD File.* This file covers the period 1957–1966, the sixth and seventh collective periods, and a limited number of records prior to 1957, primarily for substances containing fluorine and silicon. You can enter a CAS Registry Number, and the answers will be CA accession

numbers, which are the volume, column number, and column position letter of the abstracts. In this period the abstracts did not have individual numbers. You must then go to the printed copy abstracts, since no bibliographic or text information is in CAOLD.

4. *CApreviews File.* This file contains titles and bibliographic information and, for some records, the abstract and CAS Registry Numbers from approximately 100,000 articles that are in preparation for the CA file. It is updated daily, and covers journals within a week or two of receipt.

5. *CASREACT.* This is a file of over a million one-step and two million multistep organic reactions from the journal literature and patents since 1985. This may be searched by structures, reaction sites, reaction roles, and other features. You may begin with a substructure for a reactant and a substructure for a product. The answers are reaction diagrams with all reaction participants (reactants, solvents, catalysts, products) and CA abstract numbers as well as the bibliographic and indexing information and the abstracts.

6. *LCA, LREGISTRY, and LCASREACT.* These are learning files in which to become familiar with searching tactics and answers. They are small samples of the full files, and are available for low fees.

7. *Beilstein Database.* This file contains 4.5 million compounds along with their physical and chemical properties, as well as their sources and reactions. They may be found by entering a drawn structure of the compound. Substructures may be searched to find groups of compounds, and they can be searched with potential precursors in order to locate reactions. Names and literature references are given also.

8. *CJO (Chemical Journals Online).* This is a cluster of files that contains the searchable full text (not graphics) of many chemical journals. The 21 journals published by the American Chemical Society are filed back to 1982 as the CJACS file. Many others were begun in the ensuing years. These are usually searched for combinations of terms that appear adjacently or within a sentence. Answers may be displayed as short portions of the article with bibliographic data.

A search for a specific compound, for which the CA systematic name is readily found, may be efficiently done in the printed indexes. Likewise a well-defined subject that can be specified by controlled vocabulary may be found efficiently in the printed indexes. Such searching is quick and inexpensive, and may also bring unexpected related material to your attention in nearby entries in the indexes. On the other hand, online searching can provide answers that would be inaccessible from the printed indexes. Substructure searching in the Registry file can lead to compounds not found in nomenclature or molecular formula searches. The CA file contains many more searchable words than the controlled vocabulary of the indexes. Many terms lead to unusably long lists of answers, but the ability to select answers that are connected to two or more compounds or terms gives valuable short lists of answers.

STN Express® is a software package that allows you to set up your search offline. It will automatically carry out the logon procedure and upload your search, which can include structures prepared with ChemDraw®. STN Express then allows you to capture the entire session with answers electronically, and then edit and print it. Another software package called STN Mentor Laboratory™ is available on which to learn online searching without actually being online. There are also published guides on how to use the online services.

## PROBLEMS

1. Search the literature for the following compound. Give the registry number, the journal reference, the melting or boiling point, and the infrared frequency for the absorption of the carbonyl group.

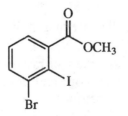

2. Find the best literature preparation of the following compound. Give the journal reference, the melting point, and a published name for the compound.

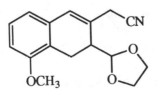

3. What is the structure of Topotecan, and what is it used for?

4. Locate a reference to the preparation of cyclooct-4-enylmethyl bromide.

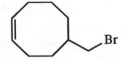

**5.** Locate a reference to a compound that is an example of the following ring system and give the CA name for that ring system.

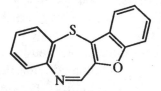

**6.** Locate a reference to the following compound where R is a substituent. Give the complete structure.

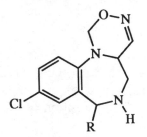

**7.** Using *Beilstein*, find the melting point of 4-bromo-3,4′-dinitrobiphenyl.

**8.** Using *Beilstein*, find the ultraviolet spectrum of biacetyl, $CH_3COCOCH_3$. Give the original reference, and list two wavelengths of maximum absorbance.

## REFERENCES

1. Antony, A. *Guide to Basic Information Sources in Chemistry*, Wiley, New York, 1979.
2. Bottle, R. T. *Use of the Chemical Literature*, 3rd ed., Butterworths, London, 1979.
3. Maizell, R. E. *How to Find Chemical Information*, 2nd ed., Wiley-Interscience, New York, 1987.
4. Skolnik, H. *The Literature Matrix of Chemistry*, Wiley-Interscience, New York, 1982.

5. Wiggins, G. *Chemical Information Sources*, McGraw-Hill, New York, 1991.
6. Schulz, H. *From CA to CAS ONLINE*, VCH Verlagsgesellschaft, Weinheim, 1988.
7. Runquist, O. *A Programmed Guide to Beilstein's Handbuch*, Burgess, Minneapolis, 1966.

# 3

# STEREOCHEMISTRY

The shapes and properties of molecules can depend not only on the order of connection of atoms but also on their arrangement in three-dimensional space. Molecules differing only in configuration are called *stereoisomers*, and are the principal subject of this chapter.[1]

## 3.1 REPRESENTATIONS

Some organic molecules such as benzene are planar as defined by the point locations of all nuclei present. These are easily represented on the planar printed page.

Most organic molecules are three-dimensional structures, best viewed and represented in solid molecular models. The necessity of using paper requires pictures that show depth, as perspective does in artwork and photography. The mere projection onto the plane of the paper, as in the shadow of a molecular model, loses the real difference between left- and right-handed structures. The best alternative on paper is a stereo pair of pictures as exemplified in Fig. I. The image on the left is for your left eye and that on the right for your right eye, a pair of views representing a 6° rotation of the molecule.

These are precisely scaled and oriented no farther apart than the separation between your eyes. You can view them through a stereopticon, or wear a pair of highly magnifying reading glasses and, with your nose close to the paper, try for a merged image. Then slowly draw your head

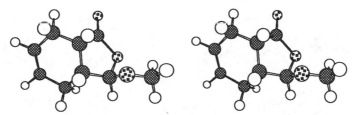

**Figure I.** A stereo pair of images of the compound produced in Eq. 27 of Chapter 6.

up from the page until it is in focus. There will appear to be three images; concentrate on the middle one, and it should look truly three-dimensional. Most commonly in journals and handwritten material we use representations where depth is portrayed via conventions instead of the pictorially obvious. These are exemplified in Fig. II for 2-butanol.

In the pictorial representation your three-dimensional cues are the front and back emergence of bondsticks on the spherically shown atoms. In the dot-and-wedge convention the group on the broad end of the wedge is defined as being above the plane of the paper, the dotted bond extends below the plane of the paper, and the line bonds are in the plane of the paper. In the abbreviated form the hydrogens on carbon are not shown but defined as completing the tetravalency of carbon.

In the Fischer projection, the center of the crossed lines is a carbon atom, and those bonds emanating from it to the side are defined as extending above the plane of the paper toward the viewer, and those extending toward the top and bottom of the page are defined as extending below the plane of the paper, away from the viewer.

Related conventions are used for portraying ring compounds as exemplified in Fig. III for (*S*,*S*)-1,2-cyclohexanediol. Many authors will draw one enantiomer of each molecule in a reaction scheme when they

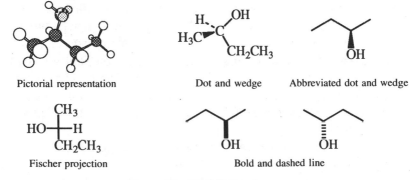

| Pictorial representation | Dot and wedge | Abbreviated dot and wedge |

| Fischer projection | Bold and dashed line |

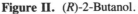

**Figure II.** (*R*)-2-Butanol.

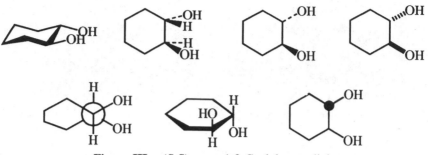

**Figure III.**   *(S,S)-trans*-1,2-Cyclohexanediol.

are actually using racemic materials. Their text should indicate this meaning.

## 3.2  VOCABULARY

A molecule or other object that is different from its own mirror image (e.g., a shoe) is *chiral*. Molecules that are identical to their mirror image are *achiral*. The conformational flexibility of molecules allows many different representations, therefore when testing a pair of structures for identity or a mirror-image relationship, the models should be flexed or the drawings redrawn to attempt a match. A left-hand fist is not the mirror image of an open right hand, yet we will refer to left and right hands as mirror images generally.

A chiral molecule and its mirror image molecule are *enantiomers*; that is, their relationship is enantiomeric. A pair of shoes is an enantiomeric pair. A *racemic mixture* or *racemate* is a combination of equal amounts of enantiomers, while a combination of unequal amounts of enantiomers is called *aracemic* or *scalemic*. A sample of a single enantiomer is termed *enantiopure*.

Most common chiral molecules contain one or more stereogenic atoms. A *stereogenic atom* is an atom bearing several groups whereamong an interchange of two groups will produce a stereoisomer (enantiomer or diastereomer) of the original. Carbon 2 in 2-butanol and carbons 1 and 2 in 1,2-cyclohexanediol are stereogenic atoms. The two possible spatial arrangements about a stereogenic atom are called *configurations*, and each one is designated (*R*) or (*S*) according to the Cahn–Ingold–Prelog system.[2]

If a molecule contains more than one stereogenic atom, there will usually be *diastereomeric* pairs. Diastereomers have the same order of connection of atoms in their structures, but one differs in spatial arrangement from the other and from the mirror image of the other in all reasonable conformations. Diastereomeric substances must, therefore, differ in many physical properties. (The term *diastereomer* is also used to relate cis and trans alkenes and cis and trans ring compounds even if they are achiral.) The term *stereoisomer* includes enantiomers and diastereomers. All three stereoisomers of 1,2-cyclohexanediol are shown in Fig. IV. The (*S,S*) and (*R,R*) isomers are mirror images and, therefore, enantiomers. The (*R,S*) isomer is different from either (*S,S*) or (*R,R*) and is, therefore, a diastereomer of each. Note that the (*R,S*) isomer is achiral despite of the presence of stereogenic atoms. Achiral molecules containing tetrahedral stereogenic atoms are termed *meso*. Meso structures may be recognized by the presence of a mirror plane within the molecule in certain conformations, such as a boat conformation in this case.

You can expect more stereoisomers for structures that contain more stereogenic atoms. For example, the $C_6$ glycopyranoses contain five stereogenic atoms (Fig. V), and there are 32 stereoisomers. Half of them are enantiomers of the other half.

Each additional stereogenic atom doubles the number of stereoisomers except where meso compounds occur, or where a polycyclic ring system prohibits some configurations. Compound **1** contains four stereogenic atoms, but there are only two stereoisomers.

**1**

Certain diastereomeric relationships are designated by prefixes derived from two carbohydrates. D-Threose **2** and D-erythrose **3** have two stereogenic atoms, and both bear an —H and an —OH group. Other molecules that differ in the analogous fashion are prefixed *threo-* and *erythro-*, as generally represented in Fig. VI.

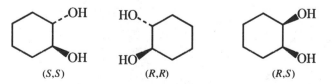

**Figure IV.** 1,2-Cyclohexanediol stereoisomers with configurational designation.

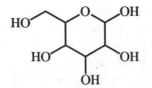

**Figure V.** The $C_6$ glycopyranoses.

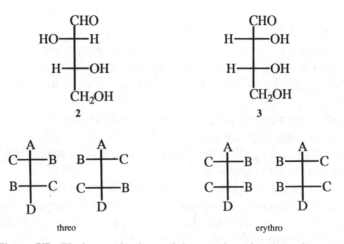

**Figure VI.** Fischer projections of threo and erythro stereoisomers.

Chirality is a property of the whole molecule, as concluded by Pasteur even before the structural theory was established, and does not require the presence of a stereogenic atom. Molecules with a twist along an axis such as allenes **(4),** spiro compounds **(5),** and exocyclic double-bonded compounds **(6)** can be chiral. Crowding of groups in a molecule may restrict rotation about single bonds or prevent planarity, again generating a twist, giving conformational enantiomerism **(7, 8).**

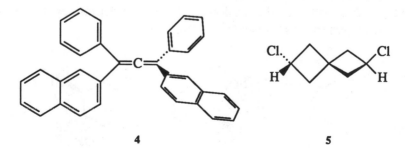

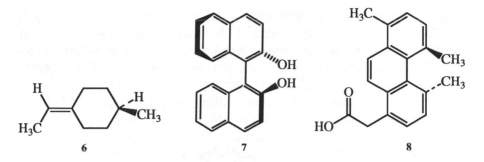

**6**            **7**            **8**

Finally, some molecules having a planar portion with one face distinguished from the other and lacking a plane of symmetry are chiral. Examples include paracyclophane **9** and *trans*-cyclooctene **10**.[3]

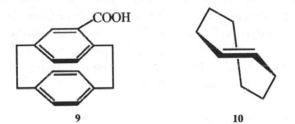

**9**                    **10**

## 3.3  PROPERTY DIFFERENCES AMONG STEREOISOMERS

Thus far we have considered structures. Structural differences should be manifest in measurable property differences in actual substances.

Diastereomers have different shapes, and the different manner in which each assembles with its own kind in the liquid or solid state* will lead to different physical properties such as density, refractive index, melting point, and boiling point. The diastereomers will also have dipolar differences, which will influence interaction with radiation, leading to different spectroscopic properties.

The difference between enantiomers is very subtle. A homogeneous sample of a pure enantiomer will have properties dependent on the intermolecular attractions in the sample, and the mirror image will be no different in that regard. Thus the melting point, boiling point, refractive index, and density of each will be identical. So also will be their spectra.

*With the lack of assembly in the gaseous state, shape is of little consequence. For example, density depends largely on molecular weight as applied in the Victor Meyer method for determining molecular weight.

Enantiomers will interact identically with achiral materials, but if each is allowed to interact with one and the same enantiomer of a chiral material, the combinations will be diastereomeric, and therefore different. That chiral material can be a solvent, adsorbant, complexing agent, or reactant. A left foot will fit well in a left shoe but not in a right shoe. The left foot–left shoe combination is diastereomeric with the left foot–right shoe combination and thus has different properties.

Ratios of enantiomers in scalemic materials can be measured using an enantiopure derivatizing or complexing agent that makes the enantiomers into diastereomeric molecules or complexes, which are distinguishable in NMR (nuclear magnetic resonance) spectra.[4] Enantiomeric primary and secondary alcohols are readily distinguished using (4R,5R)-2-chloro-4,5-dimethyl-2-oxo-1,3,2-dioxaphospholane **11**, which forms diastereomeric phosphorylated alcohols.[5] These give separate signals in $^{31}$P NMR spectra that are readily integrated. Enantiopure $\alpha$-methoxy-$\alpha$-trifluoromethylphenyacetyl chloride **12** reacts with chiral alcohols or amines to give diastereomeric esters or amides, which show separate signals in $^1$H or $^{19}$F NMR which can be measured to determine the ratio of enantiomers of the alcohol or amine.[6]

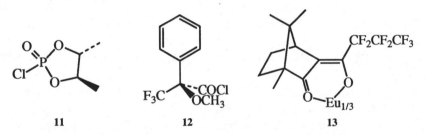

|    11    |    12    |    13    |

A chiral solvating agent such as enantiopure 2,2,2-trifluoro-1-(9-anthryl)ethanol (TFAE) combined in a nonpolar solvent with a chiral solute causes nonequivalence in the $^1$H NMR spectrum of the enantiomers of solute.[7] The solute must be capable of hydrogen bonding with the solvating agent in order to form diastereomeric complexes.

Much larger differentiations between enantiomers may be brought about with lanthanide shift reagents such as tris[3-((heptafluoropropyl)hydroxymethylene)-d-camphorato]europium [Eu(hfc)$_3$] **13**. These paramagnetic complexes associate with polar sites in molecules and cause large changes in $^1$H and $^{13}$C NMR chemical-shift values for nearby atoms. The chiral ligands in the complexes cause different shifts in enantiomeric molecules.[8] Although the shift differences are larger than those found with solvating agents, the shift reagents cause line broadening and require optimized mole ratios between reagent and the chiral substance under analysis.

It is common to express the relative amounts of enantiomers as percent enantiomeric excess (ee). This is calculated from the areas of the NMR signals for each enantiomer as in Eq. 1, where $A_L$ and $A_S$ represent the areas of the larger and smaller signals. The remaining percentage is considered as racemic. A sample of 94% ee contains 97% of one and 3% of the other enantiomer.

$$\frac{A_L - A_S}{A_L + A_S} \times 100 = \% \text{ enantiomeric excess} \qquad (1)$$

There is one direct physical measurement that allows a differentiation between enantiomers, and can be used to determine ratios of enantiomers: rotation of polarized light. Light emerging from a polarizer may be understood as follows. Each of the multitude of parallel rays is a pair of dipoles perpendicular to the line of travel that rotate as they advance, one spiraling to the right and one spiraling to the left. These dipoles alternately reinforce and cancel each other, such that their vector sum is an oscillating wave in one plane. Coming through the same polarizer, all the rays summarize in parallel planes, which means that the right and left dipole pairs all have the same phase relationship. When such a bundle of rays travels through a liquid aracemic material, the speed of one of the rotating dipoles is retarded more than the other. One enantiomer is interacting more with one spiral, while the other enantiomer will interact more with the other (mirror image) spiral. This is analogous to the selectivity of diastereomeric interactions discussed earlier. This results in a change in their phase relationship, and their vector sum is in a new plane rotated from the original. Further penetration through the substance causes further rotation of the plane. The amount of rotation is a characteristic of the substance, and is conventionally specified for a sample of density 1 g/mL with a 10-cm pathlength [$\alpha$]. The sign of the rotation is + for clockwise as viewed from the end where the light emerges, and − for counterclockwise. Of course, enantiomers give equal but opposite rotations. Measurements made with different concentrations and/or lengths may be proportioned to the specific value. Samples that rotate the plane of polarized light are called *optically active*. Racemic material is not optically active. If a sample is 75% of one enantiomer and 25% of the other, the rotation will be one-half of the maximum and that sample will be labeled 50% optically pure or 50% enantiomeric excess. Rotation measurements are commonly reported for light from a sodium lamp of wavelength 5890 Å, indicated by a subscript D. If a white light source is used, you will see various colors as one polarizer is rotated because the rotation of the plane by a sample varies with the wavelength. A graph of the rotation versus wavelength is called an *optical rotary dispersion curve.*

As mentioned before, the melting point of one enantiomer of a chiral substance will be identical to that of the other enantiomer. However, if both enantiomers are together in a sample, the melting point is likely to be different from that of a pure enantiomer. Consider rows of people shaking hands using all right hands (or all left hands). A certain fit will exist. Consider instead random shaking, including right with left. This produces a different fit. By analogy, interactions between (+) and (−) enantiomers will be different from (+) and (+), and we can expect a different melting point for the mixture. If solutions or melts of various ratios of enantiomers are cooled to produce solid, we find one of three possible behaviors:

1. The enantiomers may crystallize separately, giving a mixture of (+) and (−) crystals called a *conglomerate.* The melting point of a conglomerate is then a mixed melting point and is lower than that of pure enantiomer. Various ratios of (+) and (−) give the melting points graphed in Fig. VIIa.

2. In other cases a stronger attraction exists between opposites, and a *racemic compound* is formed. The crystals will contain both enantiomers in equal amounts. The melting point of the racemate is depressed by adding a small amount of either enantiomer (Fig. VIIb or VIIc). The racemate may melt either lower or higher than a pure enantiomer.

3. If the difference between enantiomers is small, the substance may crystallize as an ideal solid solution, where the ratio of (+) to (−) has no effect on the melting point, and the graph is flat. A few give nonideal solutions with maxima or minima.

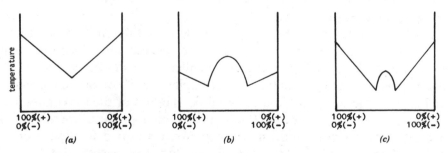

**Figure VII.** Melting-point behavior of various enantiomeric pairs.

An example of the behavior exhibited in Fig. VII*b* is found with *trans*-2-*tert*-butylcyclohexanol.[9] The pure (−) rotating enantiomer and the pure (+) rotating enantiomer melt at 50–52°C, while the racemic material melts at 84.4–85°C.

## 3.4  RESOLUTION OF ENANTIOMERS

The diastereomeric compounds or complexes discussed in Section 3.3 are used to separate racemic materials into enantiopure components, a process called *resolution*.[10] Chromatographic separation of racemates on various chiral substrates has been demonstrated.[11, 12] A particularly effective column consists of silica functionalized as in Fig. VIII, where (*R*)-phenylglycine is the chiral component.[13] The silica containing 0.70 mmol of chiral sites per gram and packed in 2 × 30-in. stainless-steel columns gave complete separation of gram quantities of various racemates eluting with 5–10% 2-propanol in hexane. One enantiomer interacting with the chiral group on the silica by hydrogen bonding, π-donor complexation, and/or dipole orientation is diastereomeric with the complex from the other enantiomer and will differ in dissociation equilibrium constant, thus eluting at different rates. Narrower columns, resolving smaller amounts with greater efficiency, have been used to measure ratios of enantiomers with ultraviolet absorption detection of the isomers in the elluate. Many columns with chiral stationary phases are commercially available.

Hydrolytic enzymes are often able to catalyze the aqueous hydrolysis of one enantiomer of a racemic ester, thus providing one enantiomer of the alcohol and the ester of the other enantiomer, readily separable by extraction or other means. Lipase Amano P, from *Pseudomonas fluorescens*, can be used more efficiently as an enantioselective catalyst for

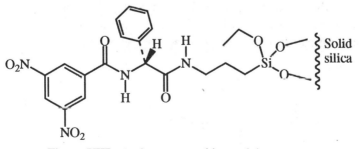

**Figure VIII.** A chromatographic resolving agent.

the esterification of chiral alcohols with acetic anhydride in benzene solution (Eq. 2), thus effecting resolution of the alcohol.[14]

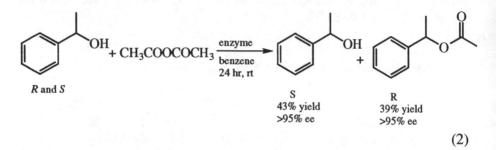

R and S

S
43% yield
>95% ee

R
39% yield
>95% ee

(2)

The resolutions described above are based on formation of diastereomeric complexes with a column stationary phase or an enzyme. The more common alternative is to bond the enantiomers covalently to a chiral resolving agent to make stable diastereomeric molecules, separate those diastereomers by chromatography or recrystallization, and then disassemble each purified diastereomer to obtain the resolved enantiomers.

Racemic *trans*-2-cyanomethylcyclohexanol was treated with (R)-(−)-1-(1-naphthyl)ethyl isocyanate to afford a pair of diastereomeric carbamates. These were separable in multigram quantities by high-pressure liquid chromatography on silica gel eluting with benzene-ether.[15] Each separate diastereomer was treated with trichlorosilane to regenerate the isocyanate resolving agent and release a pure single enantiomer of *trans*-2-cyanomethylcyclohexanol. This was hydrolyzed in a subsequent step to give the enantiomerically pure lactone.

(R)-(−)-Mandelic acid is a suitable resolving agent for isolating one enantiomer of various chiral alcohols by recrystallization of a mixture of diastereomeric esters.[16] For example, 2-octanol was esterified with (−)-mandelic acid, using *p*-toluenesulfonic acid as catalyst and benzene to remove the water azeotropically, affording the diastereomeric esters **14** in 97% yield. Recrystallization gave a 59% yield of one diastereomer. Saponification then afforded one pure enantiomer of 2-octanol. The

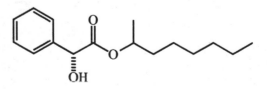

ÖH

**14**

diastereomeric purity of the ester could be determined readily by $^1$H NMR or $^{13}$C NMR where the $\alpha$-methyl group of the alcohol gave distinctly different signals in the two diastereomers.

Many carboxylic acids have been resolved by recrystallization of salts of enantiopure amines. One enantiomer and sometimes both are available this way.

The diastereomeric differentiations in the preceding two cases involve physical interactions in chromatography or crystallization. Another way is to use a chemical reaction that is fast with one diastereomer and slow with the other (kinetic resolution). This way one diastereomer is converted to a new compound, readily separable from the unchanged other diastereomer. Triisobutylaluminum converts acetals to enol ethers with this sort of selectivity. This reaction was used to resolve many unsymmetrically substituted cyclic ketones as exemplified in Eq. 3.[17]

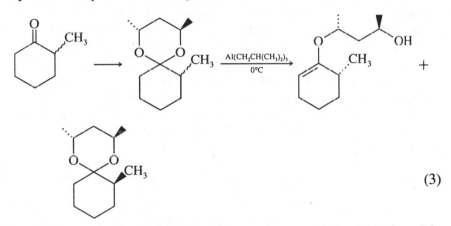

(3)

The readily separable enol ether and acetal can each be hydrolyzed in acid to give the ketone. If the reaction is carried to 35% completion, the enol ether furnishes (R)-2-methylcyclohexanone with greater than 95% enantiomeric excess. If carried to 70% completion, the remaining acetal gave (S)-2-methylcyclohexanone with greater than 95% enantiomeric excess.

Some racemates that give conglomerates may be resolved with no external diastereomeric influences.[18] Direct crystallization of individual enantiomers from saturated solutions of racemates may be localized by seeding with pure enantiomers, especially if large crystals will grow as, for example, with hydrobenzoin. A practical variation on this process, called *entrainment*, begins by enriching a solution of racemate with one enantiomer, cooling to saturation, and seeding with the one in excess. In favorable cases a crop about twice the size of the original excess is obtained. The solution then contains an excess of the other enantiomer.

More racemate is added, cooled to saturation, and seeded with the other enantiomer to gain a crop of it as large as was obtained for the first. This is then repeated indefinitely. In this way 13,000 tons of L-glutamic acid was produced annually from synthetic racemate. Many other amino acids have been resolved on a smaller scale; however, most organic racemates give racemic compounds on crystallization and are therefore unsuitable for resolution by entrainment.

## 3.5 ENANTIOSELECTIVE SYNTHESIS

In Section 3.4 pure enantiomers were obtained by separation of racemic material. Another alternative is to begin with an achiral compound and generate a single enantiomer of a chiral compound from it.[19,20] This requires a chiral influence from another component in a chemical reaction.

A chiral "template" may be temporarily attached to an achiral molecule, and then a new stereogenic center made under that influence, and finally the original "template" is removed.[21] To illustrate this, consider first the alkylation of simple ketone enolates such as cyclohexanone enolate, where a stereogenic center is formed. Without a chiral influence, the products are always racemic. If a chiral imine enolate is employed as in Eq. 4, the incoming alkylating agent is guided with high

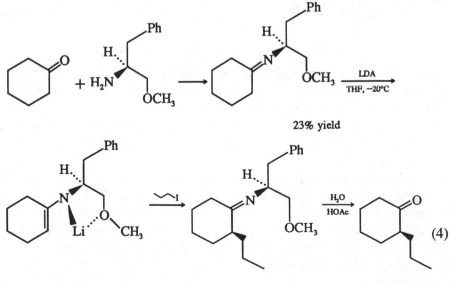

23% yield

99% (S), 76% yield

(4)

specificity to a particular face of the enolate.[22] By this procedure 12 cases on various rings and with various alkylating agents all gave the same choice of attacking face and the analogous enantiomer product. The chiral primary amine template was obtained by reduction of (*S*)-phenylalanine followed by methylation.

Templates have also been used in amide enolate alkylations,[23] conjugate additions,[24] Diels–Alder reactions,[25] Simmons–Smith reactions,[26] Claisen rearrangements,[27] and many others.

The chiral influence may be a catalyst, in which case a small amount of enantiopure material can lead to large amounts of aracemic product. Methyl acetoacetate was hydrogenated using a soluble enantiopure ruthenium complex as catalyst.[28] This gave the (*R*)-hydroxyester in 98% enantiomeric excess (Eq. 5). The catalyst is sensitive to air and must be

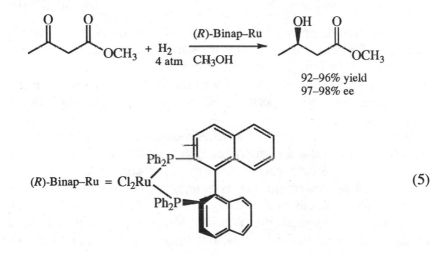

92–96% yield
97–98% ee

(5)

prepared under an inert atmosphere. The other enantiomer of the catalyst is also available; therefore, the enantiomeric alcohol is as easily prepared. This is a highly efficient multiplication of aracemic material since the substrate to catalyst mole ratio is > 1000. The *S* enantiomer of ethyl 3-hydroxybutanoate may be prepared from the ketone by reduction with bakers yeast and sucrose also.[29]

The reagent itself may be chiral. A chiral tridentate ligand prepared from (*S*)-aspartic acid was added to lithium aluminum hydride (Eq. 6). This modified hydride was then used to reduce α,β-unsaturated ketones, consistently giving the (*S*) alcohols in 57–95% isolated yield with 28–100% enantiomeric excess.[30]

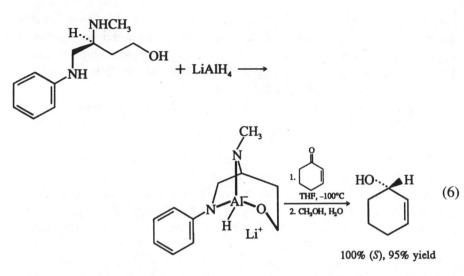

$$100\% \; (S), 95\% \; \text{yield}$$

If a resolution procedure or asymmetric synthesis gives material of perhaps 95% enantiomeric excess, simple recrystallization[31] or sublimation[32] will often give enantiopure material. Care must be taken to avoid this sort of fractionation if you are isolating a product, intending to measure the optical purity as it comes from a reaction or partial resolution.

Resolution and asymmetric synthesis depend on the energy difference between a pair of diastereomeric complexes or between crystals of diastereomers. These differences are very small compared to reaction enthalpies and are, therefore, not as generalizable. The examples selected here are some of the best in each category; there are many examples that give far less selectivity. You should not expect routine application of these techniques to new cases, but rather much trial-and-error development. On the other hand, partially resolved materials can serve very well in studies of reaction stereochemistry or in correlations of configuration.

## 3.6   REACTIONS AT A STEREOGENIC ATOM

### 3.6.1   Racemization

In molecules that contain only one tetrahedral stereogenic atom, certain conditions will lead to a loss of optical activity, eventually giving racemic material. These conditions lead to the formation of an intermediate structure where a plane of symmetry passes through the former stereogenic atom. For instance, removal of a proton from the stereogenic atom to give a carbanion allows a planar or rapidly inverting pyramidal structure to exist at that atom. The return of a proton will then be equally probable on either face, leading to either enantiomer. After sufficient time, equal amounts of enantiomers will be present, and the result is called *racemization*.

Other circumstances where a plane of symmetry may occur at an intermediate stage include nucleophilic substitution, neighboring-group participation, rearrangements and carbocation formation. 1-Bromo-ethylbenzene shows an optical half-life in solution in hexamethylphosphoric triamide–pentane of only 8 h at 27°C.[33] In this ionizing solvent, a relatively stable, flat intermediate carbocation may be responsible, or perhaps a small amount of bromide ion impurity may give nucleophilic displacement of bromide via a symmetric transition state.

### 3.6.2  Epimerization

If a proton is removed from a stereogenic atom in a molecule that contains a second, nonreacting, tetrahedral stereogenic atom, no symmetry plane is possible and the returning proton may favor one side more than the other. This may give finally unequal amounts of diastereomers, but, of course, no enantiomers. Since the second stereogenic atom is preserved, optical activity will change, but not to zero. Such a process is called *epimerization*. An example is shown in Eq. 7, where either one of the two pure diastereomers (3R,5R) or (3S,5R) was heated with a base to afford an equilibrium mixture of both where the ratio of (3R,5R) to (3S,5R) was 1.88 : 1.[34] There are many examples where the equilibrium is extreme, and one diastereomer is epimerized essentially completely to the other.

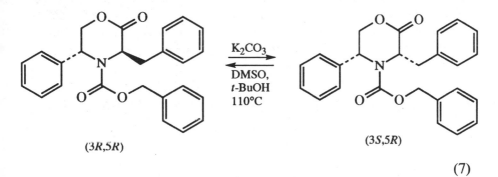

(7)

### 3.6.3  Inversion

Inversion is the replacement of a leaving group on a stereogenic atom by a new group, not in the same position but approaching from the opposite side of the stereogenic atom, causing the remaining three groups to spread through a planar condition and resume tetrahedral angles on the opposite side. One of the first examples in which inversion was known to occur in a particular step is shown in Eq. 8.[35] The overall

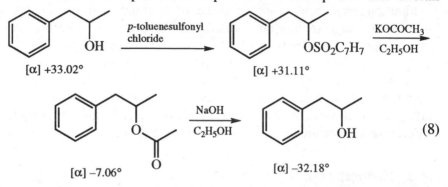

$[\alpha]$ +33.02°    $[\alpha]$ +31.11°

$[\alpha]$ −7.06°    $[\alpha]$ −32.18°    (8)

three-step process gave alcohol of inverted configuration as found by rotation measurements. The first and third step did not involve bonds to the stereogenic atom and could not give inversion. The second step must thus have given inversion. Notice that this conclusion was made without knowledge of the absolute configurations (Section 3.8). Nucleophilic substitution reactions that follow clean second-order kinetics generally give complete inversion of configuration, owing to simultaneous bond formation and breakage at the stereogenic center.

### 3.6.4   Retention

Retention of configuration occurs when an incoming group replaces a leaving group on a stereogenic atom directly (front-side) without inversion. Retention is also found when a two-step substitution occurs; that is, a temporary group arrives with inversion and is, in turn, replaced by a final group with a second inversion. The one-step front-side substitution occurs when the incoming group is attached to the leaving group and thus held to the front side in a three-membered ring transition state. Such is the case in rearrangements including the Beckmann[36] (Eq. 9),[37] Hofmann, Curtius, Schmidt, Wolff,[38] Lossen, and Baeyer–Villiger, which give retention.

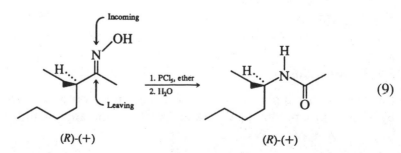

$(R)$-(+)    $(R)$-(+)    (9)

Retention was shown without use of resolved enantiomers in the Baeyer–Villiger oxidation shown in Eq. 10.[39] The oxidation gives the ace-

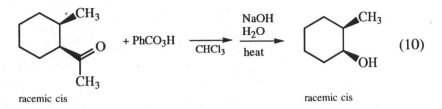

racemic cis                                                                racemic cis

tate of the alcohol which, in this case, was hydrolyzed in base. Here a second stereogenic atom is present so that retention at one site gives a diastereomer of what inversion would have given. The racemic trans starting ketone likewise gave racemic trans product. The mechanism probably includes addition of the peracid to the keto group followed by protonation and loss of benzoic acid via the cyclic transition state **15,** and finally deprotonation.

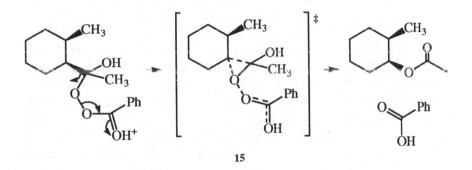

**15**

### 3.6.5  Transfer

A new tetrahedral stereogenic atom may be formed stereospecifically while an original stereogenic atom flattens. The Claisen rearrangement (Section 6.7) of aracemic allylic alcohols shows such a transfer (Eq. 11).[40] The (R)-cis alcohol gave the (S)-trans ester with 97–99% enantiospecificity. The chiral allylic alcohol and the ortho ester react to give a ketene acetal intermediate **16,** which undergoes stereospecific rearrangement via the chair–six-membered ring transition state in which the isobutyl group is equatorial. As in retention, the leaving group and the incoming group are bound together and thus disconnect from and connect to the same face of the allylic system. By close examination of the corresponding transition state, you can see that the (S)-trans allylic alcohol gives the same stereoisomeric product.

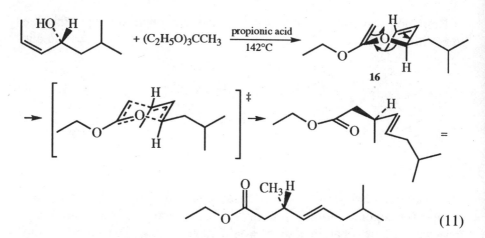

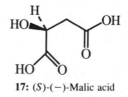

(11)

## 3.7  RELATIVE AND ABSOLUTE CONFIGURATION

If models or drawings of an enantiomeric pair are made, we can label each with an *R* or an *S* for each tetrahedral stereogenic atom. If two actual samples of the enantomeric materials are on hand, we can make a measurement of rotation direction and label each as (+) or (−). Now, which go together? Does the (+) sample have the *R* or *S* structure? Conventional X-ray structural analysis affords interatomic distances and angles, but does not provide any indication of which mirror image is present in a crystal of a chiral substance. Until the work of J. M. Bijvoet in 1951,[41] there was no way of determining this. He showed that by choosing X rays of a wavelength that excites an element in a crystal, a phase lag effect is produced that indicates which enantiomer is present. Since then, many such determinations have been done. For a particular substance, correlating the sign of rotation with the configurational designation of structure gives the *absolute configuration*. For example, in 1972[42] the cobalt(II) salt of (−)–malic acid was examined using $CuK_\alpha$ X rays to excite the Co atoms, and determined to have the structure **17**,

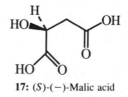

**17:** (*S*)-(−)-Malic acid

which we label *S* by the Cahn–Ingold–Prelog system. Now that this is known, the absolute configurations of many other compounds immediately become known because they have been related by synthesis at an earlier time.

For example, in 1963 (+)-2-hydroxy-3-phenylpropionic acid was ozonolyzed to the (+)-enantiomer of malic acid. This now requires that the starting acid had the $R$ configuration as shown (Eq. 12). In 1921

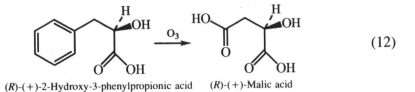

(12)

(R)-(+)-2-Hydroxy-3-phenylpropionic acid      (R)-(+)-Malic acid

(+)-2,4-dihydroxybutyric acid was oxidized to (+)-malic acid, which now requires this also to be labeled $R$.

Before the absolute determination, the relationship of these three compounds was known and useful, even though an enantiomeric picture could not be drawn with certainty. This relationship is called *relative configuration*. The early statement that (+)-2,4-dihydroxybutyric acid and (+)-malic acid and also (+)-2-hydroxy-3-phenylpropionic acid all have the "same" configuration is a statement of relative configuration. Since they do not have the identically same four groups around the stereogenic atom, the term "same" could be ambiguous, especially if several groups were modified in reactions. Therefore, a statement of relative configuration should be accompanied by a description of the reactions to be sure of the meaning of "same."

Direct conversion of compounds to others of known absolute configuration as in the preceding cases is not the only way to obtain absolute configurational assignments by relationships to established cases. Ordinary X-ray crystallography of diastereomeric substances wherein the absolute configuration at one stereogenic atom is known allows assignment of absolute configuration at the other stereogenic atoms present in the crystal structure. For example, an acid of known absolute configuration was esterified with an enantiopure alcohol of unknown absolute configuration, to give a crystalline product. The X-ray structure was determined, selecting the image in which the acid is the correct enantiomer, and reading which enantiomer of alcohol is present.[43] In another case, the absolute configuration of an acid was determined from the crystallographic structure of the amide prepared from (S)-α-methylbenzylamine.[44]

It is interesting to note that the use of configurational information such as proof of inversion in $S_N2$ reactions was made with relative configurations before absolute ones were available. In fact, the absolute configurations are not useful in themselves, except as another means of obtaining more relative configurations.[45]

Keep in mind that two compounds with the "same" configuration may have different configurational designations and/or they may have opposite signs of rotation as in Eq. 13.

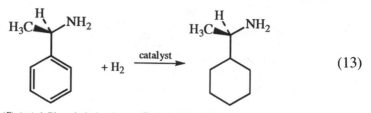

$(R)-(+)-1$-Phenylethylamine     $(R)-(-)-1$-Cyclohexylethylamine

(13)

The preceding reactions demonstrating configurational correlation did not involve bonding changes at the stereogenic carbon and are quite reliable. Many other correlations involve reactions at the stereogenic atom but with known stereochemistry such as $S_N2$ with inversion. These are reasonably reliable, also. Other correlations have been made by observing a constancy of direction of rotation in a family of compounds such as n-alkyl secondary alcohols where the S isomer is generally (+). Yet other correlations and statements of absolute configuration have been made on the basis of order of elution in chromatography[46] and on generalizations on stereospecificity in asymmetric syntheses.

Early in this chapter, absolute configurations were used in the descriptions and illustrations. All of these were established by correlations similar to those mentioned here.

An extensive, referenced, illustrated list of stereochemical correlations and absolute configurations is available,[47] and also a list of 6000 selected absolute configurations.[48]

## 3.8  TOPISM

Up to this point we have considered whole molecules differing as stereoisomers. We now turn to an atom or group (A/G) within a molecule and examine the three-dimensional shape of the environment of that A/G, within the molecule.[49] An A/G that resides in a chiral environment in a molecule is called *chirotopic*. All atoms in a chiral molecule are chirotopic, but some A/G in some achiral molecules are chirotopic, also. For example, bromochloromethane is not chiral and has no stereoisomers, but the environment of each of the hydrogen atoms is chiral (Fig. IX); therefore, those hydrogen atoms are chirotopic.

In most molecules the tetrahedral stereogenic atoms are chirotopic as well. This is not always the case as exemplified by carbon 3 in the stereoisomers of 2,3,4-trichloropentane (Fig. X). It is useful to compare

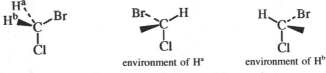

**Figure IX.** Environments in bromochloromethane.

two chirotopic A/Gs within the same molecule, that have the same bonding connectivity. We will find that they are either *enantiotopic* (have enantiomeric environments), *diastereotopic* (have environments that are more different than mirror image), or *homotopic* (have identical environments). Consider again the two hydrogens in bromochloromethane. These hydrogen atoms reside in mirror-image environments, and are thus enantiotopic to each other. There are no enantiotopic A/Gs in molecules that are chiral, but they exist in meso compounds and molecules as simple as butane.

In ordinary circumstances enantiotopic A/G exhibit identical character, but in association with chiral materials, their environments become more different than mere mirror images. This differentiation is demonstrated graphically for the general case in Fig. XI. The molecule containing enantiotopic groups can attach to a chiral substrate with all complementary groups and sites matched using one of the enantiotopic groups; however, attempting to use the other enantiotopic group fails to give a match. They are thus differentiable. Observable differentiations of this sort occur in enzyme-catalyzed reactions. For example, isobutyric acid can be hydroxylated to give enantiopure (+)-3-hydroxy-2-methyl-propionic acid by bacterial fermentation (Eq. 14).[50]

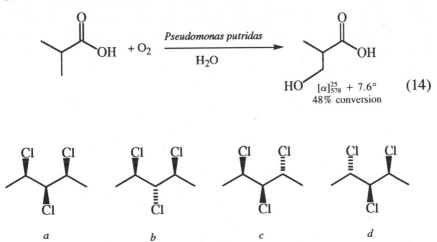

**Figure X.** The stereoisomers of 2,3,4-trichloropentane. Isomers *a* and *b* are both meso, and carbon 3 is stereogenic but not chirotopic. Isomers *c* and *d* are racemic, and carbon 3 is chirotopic but not stereogenic.

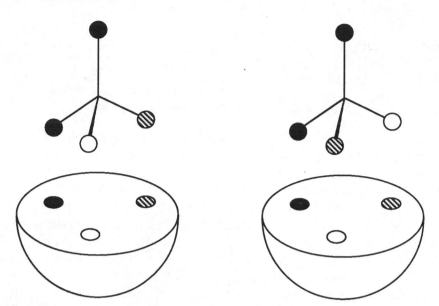

**Figure XI.** Two orientations for complexing enantiotopic groups on a chiral surface.

The two methyl groups in isobutyric acid are enantiotopic, and the chiral enzyme has selectively oxidized the same one in all the molecules.

There is a simple thought test to determine whether certain groups in a molecule are enantiotopic. Simply imagine replacing one of the two groups with an atom X. Then do likewise with the other one instead (Fig. XII). If this produces a pair of enantiomers, the groups were enantiotopic. These distinguishable enantiotopic groups may each be labeled. If the X group has a higher priority (Cahn–Ingold–Prelog) than the group it replaced but not higher than the next-higher original group, and the resulting stereogenic atom is of *R* configuration, the original enantiotopic group replaced is designated *pro-R*. If the other enantiotopic group had been replaced, the stereogenic atom would have been *S*;

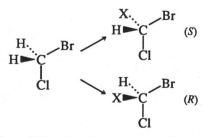

**Figure XII.** Identifying enantiotopic pairs.

therefore, the other enantiotopic group is designated *pro-S*.[49] Thus $H^a$ in Fig. IX is *pro-S* and $H^b$ is *pro-R*. This is analogous to identifying the left and right sleeves of a jacket.

Pairs of A/G that reside in diastereomeric environments are called *diastereotopic*. These, too, can be identified by the thought test. If replacing each of a pair of A/G with X produces a pair of diastereomers, the original pair of A/G are diastereotopic. Since diastereotopic A/G are not related by molecular symmetry, they may react at different rates and they will not be equivalent in NMR spectra. The $H^a$ and $H^b$ in acetaldehyde diethylacetal (Fig. XIII) have this relationship. Replacement of $H^a$ with X generates two new stereogenic atoms. Replacement of $H^b$ gives a structure diastereomeric with the first one; thus $H^a$ and $H^b$ are diastereotopic. The environments within the molecule of $H^a$ and $H^b$ are diastereomeric, and $H^a$ and $H^b$ give separate signals in the $^1H$ NMR spectrum (Chapter 10). In Fig. XIII the two ethyl groups are enantiotopic. You should build molecular models to assure yourself of these relationships and also those listed in Fig. XIII. In contrast, the diethyl acetal of formaldehyde contains no diastereotopic A/G. Diastereotopic A/G are found in most chiral molecules, meso molecules, and molecules that include an atom that bears two identical A/G and two other different A/G: Caabc.

Pairs of A/G that have identical environments and can be interchanged by rotation of the molecule or rotations within the molecule, are absolutely indistinguishable and are called *homotopic*. Substitution of X for either of them gives the same identical molecule. For example, the hydrogens of the free rotating methyl group in 1-chloro-1-bromoethane are homotopic, and the chlorine atoms in (*R,R*)-2,3-dichlorobutane are homotopic.

The two faces of a flat molecular site may be enantiotopic. If an $sp^2$ carbon has three different groups bonded to it, the faces are enantiotopic as exemplified in Fig. XIV. Looking directly at one face, if the three

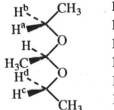

| | |
|---|---|
| | $H^a$ and $H^b$ are diastereotopic |
| | $H^c$ and $H^d$ are diastereotopic |
| | $H^c$ and $H^a$ are enantiotopic |
| | $H^b$ and $H^d$ are enantiotopic |
| | $H^a$ and $H^d$ are diastereotopic |
| | $H^b$ and $H^c$ are diastereotopic |

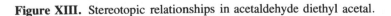

**Figure XIII.** Stereotopic relationships in acetaldehyde diethyl acetal.

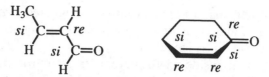

**Figure XIV.** *trans*-2-Butenal shown with the 1*si*, 2*re*, 3*si* faces toward you, and 2-cyclohexen-1-one with the 1*re*, 2*si*, 3*si* face upward.

groups give a clockwise decrease in priority, the face toward you is designated *re*. If it is turned over, the face now toward you will give counterclockwise decrease and be designated *si*.[51]

As with enantiotopic groups, enantiotopic faces are differentiated in reactions by chiral reagents or catalysts. The chiral hydride reducing agent in Eq. 6 selectively adds a hydride to the *re* face of the carbonyl carbon of $\alpha,\beta$-unsaturated ketones. A complementary hydride reducing agent with a different tridentate ligand was also prepared for selective addition of a hydride to the *si* face of these ketones.[30]

## PROBLEMS

1. How would you convert (S)-7-methyl-6-nonen-3-ol to (S)-7-methyl-6-nonen-3-ol acetate? How would you convert (S)-7-methyl-6-nonen-3-ol to (R)-7-methyl-6-nonen-3-ol acetate?[52]

2. Suppose that you needed one pure enantiomer of *exo*-bicyclo[2.2.1]heptan-2-ol. How would you prepare it from norbornene (bicyclo[2.2.1]hept-2-ene)?[16]

3. How would you carry out the following conversion to give mostly one enantiomer from the meso diester?[53]

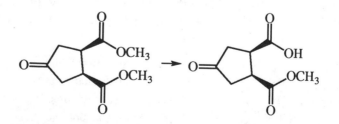

4. The following reactions were used to determine the absolute configuration of 3-hydroxy-2-methylpropanoic acid. (+)-(S)-Alanine was esterified and reduced with $LiAlH_4$ to give 2-amino-1-propanol. Phosgene then gave the 4-methyl-1,3-oxazolidin-2-one, mp 53–54°C, $[\alpha]_{578}^{20}$ − 7.8°. A sample of racemic 3-hydroxy-2-methyl-

propanoic acid was resolved to give material of rotation $[\alpha]_{578}^{20}$ − 7.6°. The resolved acid was esterified and then treated with hydrazine to give the hydrazide. Treatment of the hydrazide with nitrous acid gave a Curtius rearrangement affording the 4-methyl-1,3-oxazolidin-2-one, mp 53–54°C, $[\alpha]_{578}^{20}$ + 7.9°. Assign the absolute configuration of the acid.[54]

5. Using the information in problem 4, determine whether the *pro-R* or *pro-S* methyl group is oxidized in Eq. 14 of this chapter.

6. (+)-3-Tetradecanol is known to have the *S* configuration. Treatment of (+)-1,2-epoxytridecane with methyllithium gives (+)-3-tetradecanol. What is the absolute configuration of (+)-1,2-epoxytridecane? Draw a three-dimensional representation of it. How would you determine the absolute configuration of (−)-2-tridecanol?[55]

7. The specific rotation of the (+)-3-tetradecanol prepared in problem 4 was $[\alpha]_D^{25}$ + 6.7°. The previously reported rotation for this compound was $[\alpha]_D^{25}$ + 5.1°. What is the maximum possible enantiomeric excess in the 5.1° rotating sample? What is the maximum percent of (+)-enantiomer in the 5.1° rotating sample?

8. The epoxidation of *trans*-2-buten-1-ol with *tert*-butyl hydroperoxide catalyzed by titanium tetraisopropoxide in the presence of (+)-di-isopropyl L-tartrate gave aracemic *trans*-2,3-epoxybutanol $[\alpha]_D^{20}$ − 50.0°. Treatment of this epoxide with $(CH_3)_2CNCuLi_2$ gave some 2-methyl-1,3-butanediol, which was converted to a sulfonate ester selectively at the primary alcohol, and thence to the iodide (1-iodo-2-methyl-3-butanol) with sodium iodide in acetone. The iodo compound was reduced with $NaBH_4$ to 3-methyl-2-butanol of rotation $[\alpha]_D^{20}$ + 5.00°. Pure (*S*)-3-methyl-2-butanol is known to rotate $[\alpha]_D^{20}$ + 5.34°. Draw and name the major stereoisomer of the epoxide and give the percent of the enantiomer present with it. What would you expect if the D tartrate were used instead?[56]

9. If the following reaction occurs with inversion at both reacting stereogenic atoms, what stereoisomer(s) of the cyclopropane is (are) obtained?[57]

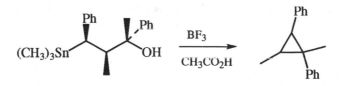

10. (+)-2,3-Dimethylsuccinic acid was converted to (+)-2-methyl-1-butanol by the following sequence of reactions. The (+)-2-methyl-

1-butanol is known to have the *R* configuration. Draw a three-dimensional representation of (+)-2,3-dimethylsuccinic acid and designate each stereogenic atom as *R* or *S*. Show how you arrived at your answer.[58]

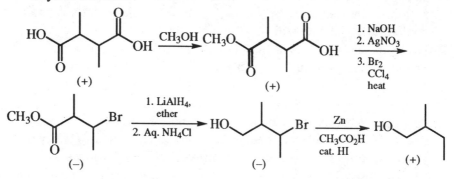

11. Draw a three-dimensional representation of 4-methylheptane. Identify a pair of enantiotopic atoms. Identify a pair of diastereotopic atoms.

12. What is the relationship between the two methoxy groups in a molecule of 3-bromobutanal dimethyl acetal?

13. What is the relationship between the two benzylic hydrogens in *meso*-2,5-diphenylhexane? What is the relationship between the two benzylic hydrogens in (*R,R*)-2,5-diphenylhexane?

14. The following steps were used to convert the (*R*)-alkynol to the (*R*)-γ,δ-unsaturated ester. How would you convert the (*R*)-alkynol into the (*S*)-γ,δ-unsaturated ester? The complex aluminum hydride is used to reduce alkynols to trans allylic alcohols.[59]

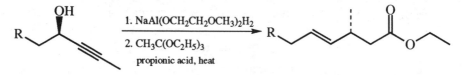

15. Draw three-dimensional representations of all the possible stereoisomeric esters of (*R*)-*O*-methyl mandelic acid derivable from all possible stereoisomers of the following alcohol.[60] How many stereoisomers from this combination are there?

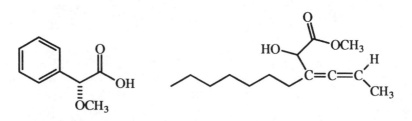

**16.** Photochemical chlorination of $(+)$-$(S)$-2-bromobutane with $t$-butyl hypochlorite at $-78°C$ gave the following, among other products:

|  | Yield | Enantiomeric Purity |
|---|---|---|
| *erythro*-2-Bromo-3-chlorobutane | 20% | Enantiopure |
| 2-Bromo-2-chlorobutane | 53% | Racemic |
| *threo*-2-Bromo-3-chlorobutane | 6% | Racemic |
| 3-Bromo-1-chlorobutane | 4% | Enantiopure |
| 1-Bromo-2-chlorobutane | 3% | Enantiopure |

Draw three-dimensional representations of each of these products, and explain in terms of intermediates why three of these are enantiopure and two are not.[61] (*Hint*: Consider the 3% yield product first.)

**17.** The two stereoisomers of 3-methyl-2,4-dibromopentane shown below were cyclized by treatment with zinc in 1-propanol–water. The products were analyzed for the ratio of *cis*- to *trans*-trimethylcyclopropanes; the results are as follows:

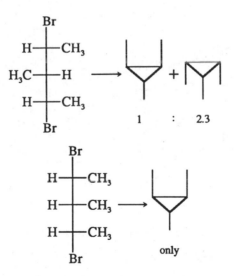

The overall stereochemical possibilities are retention at both sites, retention at one and inversion at the other site, and inversion at both sites. Considering all the results, what is the overall stereochemistry for the process or processes that give(s) the major product from the first isomer? What is the overall stereochemistry for the formation of the minor product from the first isomer? Explain with drawings how you reached your conclusions. Molecular models may be helpful.[62]

**18.** In terms of *re* and *si*, what face of the double bond received the hydrogen atoms in Eq. 5 of this chapter?

**19.** The following two sequences of reactions were carried out. Why does one give material with deuterium α to the benzene ring and the other no deuterium α to the benzene ring?[63]

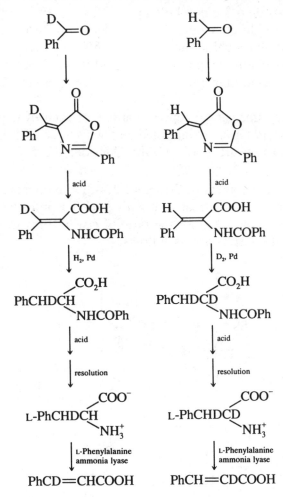

**20.** A carboxylic acid of unknown absolute configuration was converted to an amide of (*S*)-1-phenylethylamine. Conventional X-ray crystallography gave a structure for the amide that might be either of the following mirror images. What is the absolute configuration of the acid at each stereogenic atom?[44]

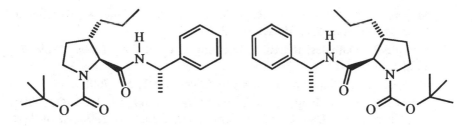

# REFERENCES

1. Eliel, E. L. *Elements of Stereochemistry*, Wiley, New York, 1969; Bassendale, A. *The Third Dimension in Organic Chemistry*, Wiley, New York, 1984.

2. Cahn, R. S.; Ingold, C. K.; Prelog, V. *Angew. Chem. Internatl. Ed.* **1966,** *5*, 385; Cahn, R. S. *J. Chem. Educ.* **1964,** *41*, 116.

3. Cope, A. C.; Banholzer, K.; Jones, F. N.; Keller, H. *J. Am. Chem. Soc.* **1966,** *88*, 4700.

4. Parker, D. *Chem. Reviews* **1991,** *91*, 1441.

5. Anderson, R. C.; Shapiro, M. J. *J. Org. Chem.* **1984,** *49*, 1304.

6. Dale, J. A.; Dull, D. L.; Mosher, H. S. *J. Org. Chem.* **1969,** *34*, 2543.

7. Pirkle, W. H.; Hoover, D. J. *Topics Stereochem.*, **1982,** *13*, 263.

8. Sullivan, G. R. *Topics in Stereochem.* **1978,** *10*, 287.

9. Djerassi, C.; Hart, P. A.; Warawa, E. J. *J. Am. Chem. Soc.* **1964,** *86*, 78.

10. Jacques, J.; Collet, A.; Wilen, S. H. *Enantiomers, Racemates and Resolutions*, Wiley, New York, 1981.

11. Blaschke, G. *Angew. Chem. Internatl. Ed.*, **1980,** *19*, 13 (review).

12. Pirkle, W. H.; Pochapsky, T. C. *Chem. Reviews* **1989,** *89*, 347.

13. Pirkle, W. H.; Welch, C. J. *J. Org. Chem.* **1984,** *49*, 138; Pirkle, W. H.; Finn, J. M., *J. Org. Chem.* **1982,** *47*, 4037.

14. Bianchi, D.; Cesti, P.; Battistel, E. *J. Org. Chem.* **1988,** *53*, 5531.

15. Pirkle, W. H.; Adams, P. E. *J. Org. Chem.* **1980,** *45*, 4111.

16. Whitesell, J. K.; Reynolds, D. *J. Org. Chem.* **1983,** *48*, 3548.

17. Mori, A.; Yamamoto, H. *J. Org. Chem.* **1985,** *50*, 5444.

18. Collet, A.; Brienne, M. J.; Jacques, J. *Chem. Rev.* **1980,** *80*, 216.

19. Valentine, D., Jr.; Scott, J. W. *Synthesis*, **1978,** 329 (review).

20. Morrison, J. D., Ed. *Asymmetric Synthesis*, Vols. 1–5, Academic Press, New York, 1983–1985.

21. Meyers, A. I. In *Asymmetric Reactions and Processes in Chemistry*, Eliel, E. L.; Otsuka, S., Eds. (ACS Symposium Series, Vol. 185), American Chemical Society, Washington, DC, 1982, p. 83.

22. Meyers, A. I.; Williams, D. R.; Erickson, G. W.; White, S.; Druelinger, M. *J. Am. Chem. Soc.* **1981,** *103*, 3081.

23. Herold, P.; Duthaler, R.; Rihs, G.; Angst, C. *J. Org. Chem.* **1989,** *54*, 1178.

24. Meyers, A. I.; Shipman, M. *J. Org. Chem.* **1991,** *56*, 7098.

25. Feringa, B. L.; de Jong, J. C. *J. Org. Chem.* **1988,** *53*, 1125; Oppolzer, W.; Chapuis, C.; Dao, G. M.; Reichlin, D.; Godel, T. *Tetrahedron Lett.* **1982,** *23*, 4781; Helmchen, G.; Schmierer, R. *Angew Chem. Internatl. Ed.* **1981,** *20*, 205. Review: Oppolzer, W. *Angew. Chem. Internatl. Ed.* **1984,** *23*, 876–889.

26. Mash, E. A.; Torok, D. S. *J. Org. Chem.* **1989,** *54*, 251.

27. Kallmerten, J.; Gould, T. J. *J. Org. Chem.* **1986,** *51*, 1152.

28. Kitamura, M.; Tokunaga, M.; Ohkuma, T.; Noyori, R. *Org. Synth.* **1992,** *71*, 1.

29. Seebach, D.; Sutter, M. A.; Weber, R. H.; Züger, M. F. *Org. Synth.* Coll. Vol. VII, **1990,** 215.

30. Sato, T.; Goth, Y.; Wakabayashi, Y.; Fujisawa, T. *Tetrahedron Lett.* **1983,** *24*, 4123.

31. Denis, J.-N.; Correa, A.; Greene, A. E. *J. Org. Chem.* **1990,** *55*, 1958.

32. Garin, D. L.; Greco, D. J. C.; Kelley, L. *J. Org. Chem.* **1977,** *42*, 1249.

33. Hutchins, R. O.; Masilamani, D.; Maryanoff, C. A. *J. Org. Chem.* **1976,** *41*, 1071.

34. Dellaria, Jr., J. F.; Santarsiero, B. D. *J. Org. Chem.* **1989,** *54*, 3916.

35. Phillips, H. *J. Chem. Soc.* **1923,** *123*, 44.

36. Gawley, R. E. *Org. React.* **1988,** *35*, 1.

37. Kenyon, J.; Young, D. P. *J. Chem. Soc.*, **1941,** 263.

38. Clark, R. D. *Synth. Commun.* **1979,** *9*, 325.

39. Turner, R. B. *J. Am. Chem. Soc.* **1950,** *72*, 878.

40. Chan, K.-K.; Cohen, N.; De Noble, J. P.; Specian, A. C., Jr.; Saucy, G. *J. Org. Chem.* **1976,** *41*, 3497.

41. Bijvoet, J. M.; Peerdeman, A. F.; van Bommel, A. J. *Nature* **1951,** *168*, 271; Trommel, J.; Bijvoet, J. M. *Acta Cryst.* **1954,** *7*, 703; Bijvoet, J. M. *Endeavour* **1955,** *14*, 71.

42. Kryger, L.; Rasmussen, S. E. *Acta Chem. Scand.* **1972,** *26*, 2349.

43. Dung, J.-S.; Armstrong, R. W.; Anderson, O. P.; Williams, R. M. *J. Org. Chem.* **1983,** *48*, 3592; Willard, P. G. *J. Org. Chem.* **1991,** *56*, 485.

44. Chung, J. Y. L.; Wasicak, J. T.; Arnold, W. A.; May, C. S.; Nadzan, A. M.; Holladay, M. W. *J. Org. Chem.* **1990,** *55*, 270.

45. Fiaud, J. C.; Kagan, H. B. *Determination of Configurations by Chemical Methods*, Georg Thieme, Stuttgart, 1977, p. 2.

46. Doolittle, R. E.; Heath, R. R. *J. Org. Chem.* **1984,** *49*, 5041.

47. Klyne, W.; Buckingham, J. *Atlas of Stereochemistry*, 2nd ed., Vols. 1, 2, Oxford University Press, New York, 1978.

48. Jacques, J. *Absolute Configuration of 6000 Selected Compounds with One Asymmetric Carbon Atom*, Georg Thieme, Stuttgart, 1977.

49. Mislow, K.; Raban, M. *Topics Stereochem.* **1967,** *1*, 1; Mislow, K.; Siegel, J. *J. Am. Chem. Soc.* **1984,** *106*, 3319.

50. Goodhue, C. T.; Schaeffer, J. R. *Biotechnol. Bioeng.* **1971,** *13*, 203.

51. Hanson, K. R. *J. Am. Chem. Soc.* **1966,** *88*, 2731.

52. Johnston, B. D.; Oehlschlager, A. C. *J. Org. Chem.* **1986,** *51*, 760.

53. Gais, H.-J.; Bülow, G.; Zatorski, A.; Jentsch, M.; Maidonis, P.; Hemmerle, H. *J. Org. Chem.* **1989,** *54*, 5115.

54. Retey, J.; Lynen, F. *Biochem. Biophys. Res. Commun.* **1964,** *16*, 358.

55. Coke, J. L.; Richon, A. B. *J. Org. Chem.* **1976,** *41*, 3516.

56. White, J. D.; Theramongkol, P.; Kuroda, C.; Engebrecht, J. R. *J. Org. Chem.* **1988,** *53*, 5909.

57. Fleming, I.; Urch, C. J. *J. Organomet. Chem.* **1985,** *285*, 173.

58. Carnmalm, B. *Arkiv för Kemi* **1960,** *15*, 215.

59. Chan, K.-K.; Specian, A. C., Jr.; Saucy, G. *J. Org. Chem.* **1978,** *43*, 3435.

60. Marshall, J. A.; Wang, X. *J. Org. Chem.* **1990,** *55*, 2995.

61. Skell, P. S.; Pavlis, R. R.; Lewis, D. C.; Shea, K. J. *J. Am. Chem. Soc.* **1973,** *95*, 6735.

62. Applequist, D. E.; Pfohl, W. F. *J. Org. Chem.* **1978,** *43*, 867.

63. Kirby, G. W.; Michael, J. *Chem. Commun.* **1971,** 415.

# 4

# FUNCTIONAL GROUP TRANSFORMATIONS

A chemist who undertakes the synthesis of an organic compound of some complexity must consider three aspects: (1) the synthesis of the functional groups in the final molecule plus those needed at intermediate stages,[1] (2) the formation of carbon–carbon bonds to develop larger molecules, and (3) the strategy of selecting starting materials and intermediate goals. A chapter is devoted to each; the first is concerned with functional groups.

The current practical alternatives for preparing each functional group include many classical reactions with relatively known mechanisms, plus many modern ones with complex or often unknown mechanisms. The introductory texts favor conceptually simple methods applied to small monofunctional molecules. Most of those synthetic products are commercially available; therefore, more generalizable and selective methods are chosen here. There are, of course, many more methods, with advantages for particular circumstances, besides these selections. The more common functional groups are covered. The many less common ones may be found in the review literature.[1] *The functional group syntheses that involve joining carbon atoms are presented in Chapter 5.*

## 4.1  CARBOXYLIC ACIDS AND RELATED DERIVATIVES

The high oxidation state of carbon in which there are three bonds to electronegative atoms is the characteristic of carboxylic acids and the related acid chlorides, anhydrides, esters, ortho esters, amides, and nitriles. The transformations may involve oxidation from hydrocarbons or other partially oxidized substrates or exchange among the various electronegative atoms on the carbon.

### 4.1.1 Carboxylic Acids

Benzylic sites containing at least one hydrogen in hydrocarbons may be oxidized to the carboxylic acid state by using strong agents including dichromate, permanganate, and nitric acid. In polyalkyl benzenes, selectivity with moderate yields may be obtained using aqueous nitric acid. The order of ease of oxidation of some alkyl groups is isopropyl > ethyl > methyl >> *tert*-butyl.[2] For example, *p*-cymene was converted to *p*-toluic acid as in Eq. 1.[3]

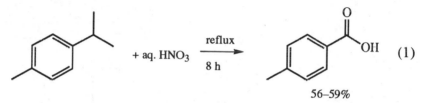

56–59%

Carbons that are already partially oxidized such as alkenes, primary alcohols, aldehydes, and methyl ketones are more readily raised to the carboxylic acid oxidation state. Appropriately substituted alkenes may be cleaved by using ozone followed by hydrogen peroxide to give carboxylic acids. A convenient alternative is the combination of sodium periodate and a catalytic amount of permanganate (Eq. 2).[4] The per-

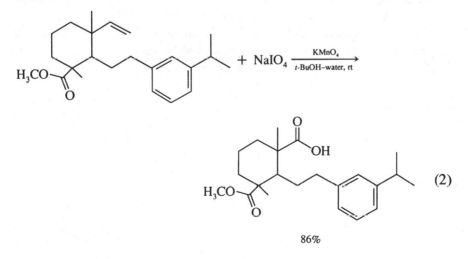

86%

manganate oxidizes the alkene to the glycol, which is then cleaved by the periodate. The periodate also regenerates the permanganate. Aqueous potassium permanganate will oxidize alkenes rapidly to the acids if a small amount of trioctylmethylammonium chloride is present as a phase-transfer catalyst (Section 9.6). In this way 1-decene was converted to nonanoic acid in 91% yield in 3 minutes.[5]

Primary alcohols are oxidized by the easily prepared pyridinium di-chromate in $DMF^{6}$ (Eq. 3)[7] or by Jones reagent in acetone (Eq. 4).[8] Secondary alcohols are also oxidized to ketones under these conditions.

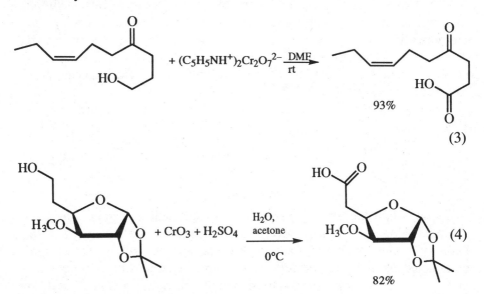

(3)

(4)

Potassium permanganate in aqueous NaOH will oxidize primary alcohols but will not be selective, attacking alkene sites as well.

Aldehydes are more readily oxidized than alcohols and thus react with the reagents given above. Nonconjugated aldehydes give acids in good yield with pyridinium dichromate in DMF. Where selectivity is needed, very mild reagents such as freshly precipitated silver oxide[9] or sodium chlorite with hydrogen peroxide (Eq. 5)[10] serve well. The hydrogen peroxide is present to destroy the hypochlorite formed in the oxidation of the aldehyde.

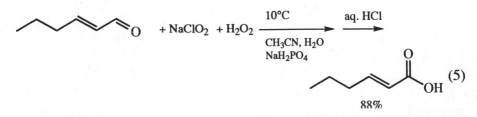

(5)

Within the same oxidation level, any of the acid derivatives may be hydrolyzed with aqueous acid or base, leading ultimately to the acid or the salt thereof. Nitrile hydrolysis is particularly difficult, requiring prolonged heating in water–ethylene glycol (Eq. 6).[11]

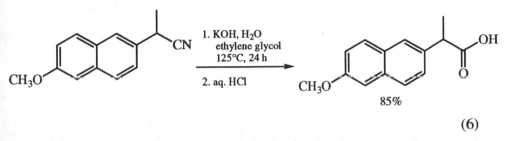

$$(6)$$

There are many syntheses of acids where a carbon–carbon bond is formed such as carbonation of Grignard reagents, malonic ester alkylation, and Reformatsky reactions. Some are covered in Chapter 5, and you should review others in the introductory text.

### 4.1.2  Carboxylic Esters

Carboxylic acids may be converted to esters directly by using a primary or secondary alcohol and a small amount of strong acid catalyst. This is a reversible equilibrium, where ester formation is favored by using excess of the alcohol or by removing the water produced. Azeotropic distillation of the water or consumption of the water by concurrent hydrolysis of an acetal are usually effective.

The highly reactive acid chlorides and anhydrides give esters irreversibly. Acetate esters of complex alcohols are routinely prepared by treating with acetic anhydride and pyridine.

Several methods are available that do not begin with alcohols. The sodium or potassium salts of carboxylic acids are sufficiently nucleophilic to displace primary iodides (Eq. 7).[12]

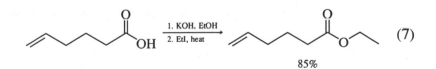

$$(7)$$

Under neutral conditions a carboxylic acid will react with diazomethane in ether to give nitrogen gas plus the methyl ester in high yield and purity (Eq. 8).[13] This is ordinarily used on a small scale because diazomethane is volatile, toxic and explosive.

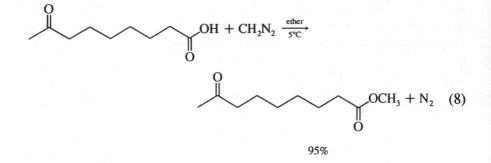

$$95\%$$

Ketones may be oxidized to esters by peracids or hydrogen peroxide, a process known as the *Baeyer–Villiger oxidation*. Unsymmetric ketones are oxidized selectively at the more substituted $\alpha$ carbon, and that carbon migrates to oxygen with retention of configuration. Trifluoroperacetic acid generated *in situ* gave the double example in Eq. 9.[14] Cyclic ketones afford lactones (Eq. 10).[15]

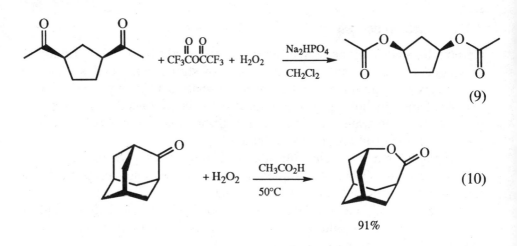

$$(9)$$

$$91\%$$

$$(10)$$

Enol esters may be prepared from ketones by reaction with an anhydride or by exchange with isopropenyl acetate under acidic catalysis (Eq. 11).[16]

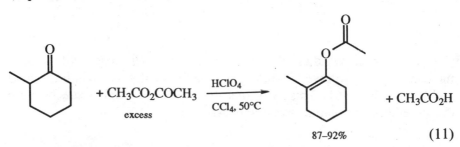

$$87\text{--}92\%$$

$$(11)$$

### 4.1.3  Carboxylic Amides

Heating a carboxylic acid with ammonia or urea gives a carboxamide. For example, heptanoic acid plus urea at 140–180°C gives heptanamide in 75% yield plus $CO_2$ and $H_2O$.[17] The highly reactive acid halides and anhydrides combine with ammonia or primary or secondary amines to give amides at ordinary temperatures. Esters will react slowly with ammonia at room temperature (Eq. 12).[18] Higher boiling amines may be used if the alcohol is removed continuously by distillation.

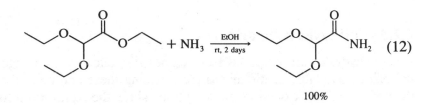

$$(12)$$

100%

Nitriles may be hydrated to amides by using acid or base catalysis and vigorous heating. The hydration may be accomplished without heating if the highly nucleophilic hydroperoxide ion is used. This is facilitated by phase-transfer catalysis as shown in Eq. 13.[19] It can also be

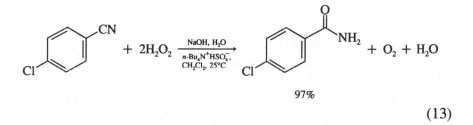

97%

$$(13)$$

done under neutral conditions using colloidal black copper catalyst (from $NaBH_4$ reduction of $CuSO_4$) at 90°C.[20] With this catalyst, the sensitive acrylonitrile was converted to acrylamide in 89% yield.

### 4.1.4  Carboxylic Acid Halides

Acid chlorides are commonly made from acids by exchange with an excess of thionyl chloride or oxalyl chloride. Brief heating gives the acid chloride plus gaseous by-products (Eq. 14).[21] A trace of dimethyl formamide accelerates this process.[22] Phosphorus tri- and pentachlorides are used similarly. The acid bromides are made with phosphorus tribromide or oxalyl bromide.[21]

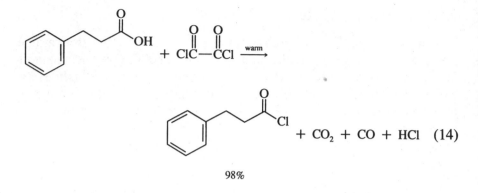

98%

## 4.1.5  Carboxylic Anhydrides

Most anhydrides are prepared from carboxylic acids by exchange with the readily available acetic anhydride. Heating these and then distilling the acetic acid and excess acetic anhydride shifts the equilibrium toward the higher-boiling product (Eq. 15).[23] Five- and six-membered cyclic anhydrides usually form simply on heating the dicarboxylic acid to about 120°C.

$$\underset{}{Ph_2CHCOH} + CH_3COCCH_3 \rightleftharpoons Ph_2CHCOCCHPh_2 + CH_3COH \quad (15)$$

90–92%

Unsymmetric anhydrides that react selectively on one side are useful. Although formic anhydride is unstable above $-40°C$, acetic formic anhydride can be prepared by stirring sodium formate with acetyl chloride in ether (64% yield, bp 27–28°C).[24] It is useful for the formylation of alcohols and amines. A stable solid formylating agent is the mixed anhydride prepared in 89% yield from p-methoxybenzoyl chloride and sodium formate catalyzed by a polymeric pyridine oxide.[25] Ethyl chloroformate gives mixed anhydrides with various carboxylic acids which are then susceptible to nucleophilic substitution at the carboxylic carbonyl carbon.

## 4.1.6  Nitriles

Nitriles may be prepared by dehydration of amides. Phosphorus pentoxide, and various acid chloride–base combinations have been used at elevated temperatures but it can be done readily at 0°C to room temperature with a Vilsmeier reagent.[26,27] Oxalyl chloride plus dimethylformamide in acetonitrile gives a precipitate of the reagent, an iminium salt, which is used as shown in Eq. 16.[26] Aldoximes may likewise be

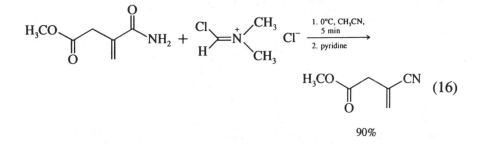

90%

dehydrated by using chlorosulfonyl isocyanate–triethylamine,[28] acetic anhydride–pyridine, or other combinations.[29] Nitriles are also commonly made by displacements with cyanide ion (Section 5.1.1).

### 4.1.7 Ortho Esters

Ortho esters are acid derivatives in which the carboxyl carbon is $sp^3$-hybridized;[30] however, most cannot be made from carboxylic acids. They are usually made by a two-stage alcoholysis of nitriles. Treatment of a nitrile with anhydrous hydrogen chloride in an alcohol gives the hydrochloride of an imidic ester. Treatment of this with an alcohol in a separate step (Eq. 17)[31] leads to the ortho esters.

$$C_2H_5CN + C_2H_5OH + HCl \xrightarrow[0°C]{} C_2H_5\overset{\overset{+}{N}H_2\;Cl^-}{\underset{\|}{C}}-OC_2H_5 \xrightarrow[Et_2O]{C_2H_5OH} C_2H_5C(OC_2H_5)_3$$

85–95%                    75–78%

$$(17)$$

The ortho formates and ortho benzoates are made from chloroform or trichloromethyl compounds by reaction with sodium alkoxides.[32]

The alcohol parts of ortho esters may be exchanged under acidic conditions to give new ortho esters, especially where the incoming alcohol is a diol. This reaction is important in some Claisen rearrangements (Section 6.7). In contrast, the acid portion cannot be exchanged; that is, an acid cannot be converted to an ortho ester directly by trans-esterification. A few acids such as chloroacetic acid may be converted to bicyclic ortho esters by reaction with a triol with azeotropic removal of water. A general route to bicyclic ortho esters begins with 3-methyl-3-hydroxymethyloxetane as shown in Eq. 18.[33] The oxetane is prepared from neopentanetriol and diethyl carbonate.

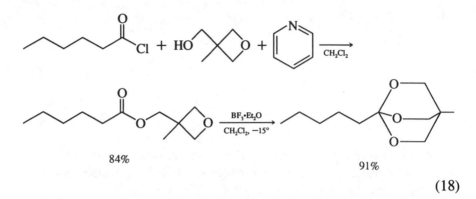

84%

91%

$$(18)$$

## 4.2  ALDEHYDES, KETONES, AND DERIVATIVES

The intermediate oxidation state of carbon in which there are two bonds to electronegative atoms is attained by reduction of acid derivatives or oxidation of alcohols and hydrocarbons. Interconversions at the same oxidation level such as hydration of alkynes and hydrolysis of vinyl halides are also valuable.

### 4.2.1  Aldehydes

The reduction of acid derivatives to aldehydes requires control because aldehydes are more easily reduced than the acid derivatives. The palladium-catalyzed hydrogenation of acid chlorides in the presence of 2,6-dimethylpyridine,[34] a modification of the Rosenmund reduction, shows this selectivity. If alkene sites are present, palladium on barium sulfate will leave them unchanged (Eq. 19); otherwise palladium on carbon is suitable. Aroyl halides require higher temperatures and quinoline-$S$-poisoned catalyst. The reaction is carried out at 1–4 atm of hydrogen pressure, monitoring gas uptake. Acid chlorides may also be reduced with sodium borohydride and $CdCl_2 \cdot 1.5$ DMF.[35]

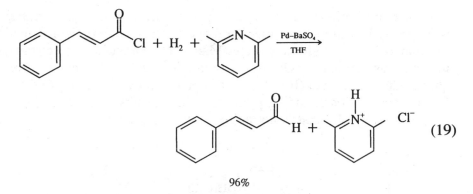

(19)

96%

Nitriles and esters, especially lactones, may be reduced to aldehydes or hemiacetals by using diisobutylaluminum hydride (Eqs. 20, 21)[36,37] or various alkoxyaluminum hydrides such as $NaAlH_2(OC_2H_4OCH_3)_2$.[38] Any free aldehyde function already present will be reduced to the alcohol even faster. The aldehyde product in Eq. 21 is in equilibrium with the hemiacetal.

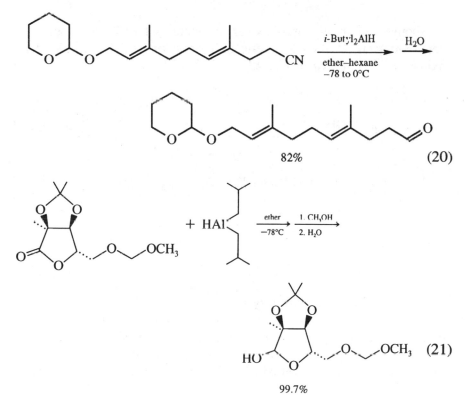

82%    (20)

99.7%

Oxidation of primary alcohols can give aldehydes, and again control is necessary because simple oxidizing agents such as chromic acid will carry on to the carboxylic acid stage. Many $Cr^{6+}$ complexes are selective for this transformation. Of these, pyridinium chlorochromate is frequently the best choice (Eq. 22).[39] It is prepared by dissolving $CrO_3$ in 6 $M$ aqueous HCl and adding pyridine, which results in a yellow filterable, air-stable solid that is not appreciably hygroscopic.

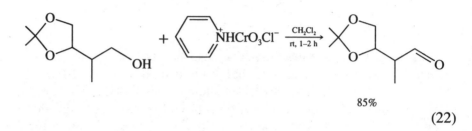

85%

(22)

Another commonly used oxidizing agent is DMSO together with a dehydrating agent such as acetic anhydride or oxalyl chloride (Swern oxidation)[40] as shown in Eq. 23.[41]

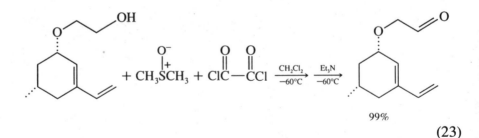

99%

(23)

Allylic and benzylic primary and secondary alcohols are more easily oxidized, and a number of reagents selective for these are in use, including freshly precipitated manganese dioxide, silver carbonate, dichlorodicyanoquinone, and potassium ferrate. 4-(Dimethylamino)pyridinium chlorochromate is mild and selective as demonstrated in Eq. 24.[42]

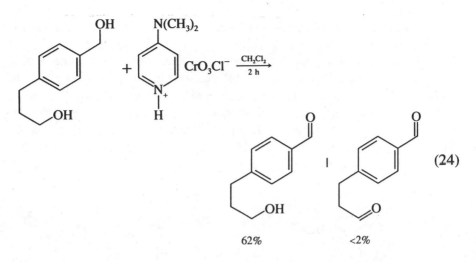

(24)

62%            <2%

Oxidative cleavage of appropriate alkenes can give aldehydes. Where ozone is used, the intermediate ozonides have more oxidizing power that can oxidize the desired aldehydes to carboxylic acids during hydrolysis. To avoid this interference, dimethyl sulfide is added as a reducing agent[43] as exemplified in Eq. 25.[44] The same overall result can

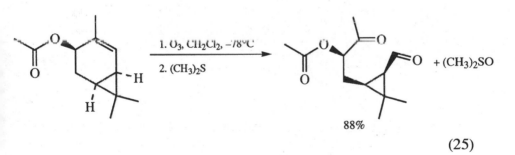

(25)

be attained by using sodium periodate with a catalytic amount of osmium tetroxide. The $OsO_4$ gives the glycol, which the periodate cleaves to the aldehyde. The periodate also regenerates the $OsO_4$.

## 4.2.2  Ketones

Ketones are far less susceptible to oxidation than aldehydes and are readily prepared by oxidation of appropriately substituted alkenes and secondary alcohols. The conditions given in Sections 4.1.1 and 4.2.1

for oxidation of alkenes to acids or aldehydes are applicable for ketones as well, as already shown in Eq. 25.

The oxidation of secondary alcohols is often done by adding $CrO_3$ and $H_2SO_4$ in water (Jones reagent)[45] to a solution of the alcohol in acetone. If an excess is avoided, alkene sites are untouched. An inexpensive, high-yielding reagent is aqueous sodium hypochlorite in acetic acid.[46,47] This gave 2-octanone from the alcohol in 96% yield. Secondary alcohols usually undergo oxidation faster than primary ones, and the selectivity can be high as with the diol in Eq. 26.

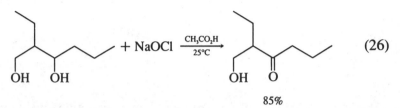

$$+ NaOCl \xrightarrow[25°C]{CH_3CO_2H} \qquad (26)$$

85%

At the same oxidation level, vinyl halides may be hydrolyzed to ketones. This may be done in cold concentrated sulfuric acid, but many polyfunctional compounds undergo further reactions in this medium. A mild alternative is treatment with mercuric salts in organic solvents followed by demercuration with dilute acid (Eq. 27).[48] This process is

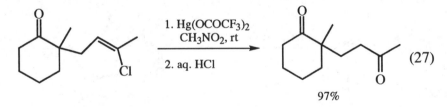

1. Hg(OCOCF₃)₂
   CH₃NO₂, rt

2. aq. HCl

$$(27)$$

97%

valuable in synthesis because alkylation of carbanions, including enolates, with 2,3-dihalopropene or 1,3-dihalo-2-butene readily provides such vinyl halides (Section 6.3).

### 4.2.3 Imines and Enamines

Imines, the nitrogen analogs of ketones and aldehydes, are commonly prepared using primary amines and dehydrating conditions[49] as exemplified in Eq. 28, where the water was removed as a benzene azeotrope.[50]

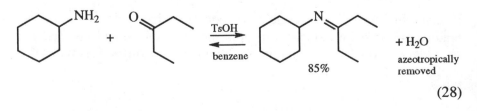

$$+ H_2O$$

azeotropically
removed

85%

$$(28)$$

If a secondary amine is used with a ketone or aldehyde, elimination cannot occur between the nitrogen and the former carbonyl carbon. Elimination then occurs between the carbonyl carbon and an $\alpha$ carbon (Eq. 29).[51] This product is called an enamine.[52] Pyrrolidine, piperidine,

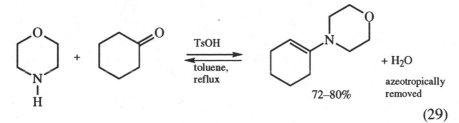

$$(29)$$

and morpholine are commonly used. Enamines and enolate anions from imines are useful in carbon–carbon bond formation (Sections 3.5 and 6.3).

### 4.2.4  Acetals

Acetals[53] are derivatives of aldehydes and ketones wherein the oxidation level remains the same but the hybridization of the carbon changes to $sp^3$. This renders the former carbonyl carbon unattractive to nucleophiles and is therefore a temporary protecting device.

Aldehydes are converted to acetals by treating with excess alcohol and an acid catalyst. The reaction is reversible, and the equilibrium is driven toward the acetal by the excess alcohol or by removal of the water as it is produced. If a 1,2- or 1,3-diol is used, the cyclic acetal forms readily, even exothermally, but is sometimes difficult to remove.

The greater steric hindrance in ketones makes acetal formation more difficult; therefore, ethylene glycol or other diols are used and the water is removed azeotropically with solvents such as toluene (Eq. 30).[54]

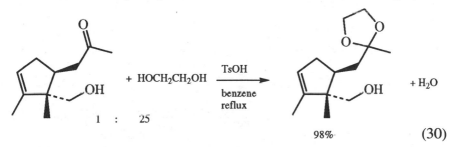

$$(30)$$

The water may alternatively be consumed *in situ* by including an ortho ester such as triethyl orthoformate, which becomes hydrolyzed to ester and alcohol during the acetalization. Trans acetalization is also useful. Treatment of a ketone with excess 2-ethyl-2-methyl-1,3-dioxolane in acid gives 2-butanone and the new ketal (Eq. 31).[55] In some cases a smaller excess of dioxolane is used and the reaction is driven by distilling the 2-butanone as it is formed.

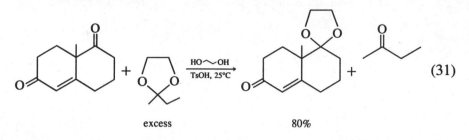

excess                                80%

Acetals can be used to protect alcohols, also. In these cases an enol ether such as ethyl vinyl ether or dihydropyran is used in place of the aldehyde as in Eq. 32. Under basic conditions an α-haloether will convert an alcohol to an acetal as in Eq. 33.[56]

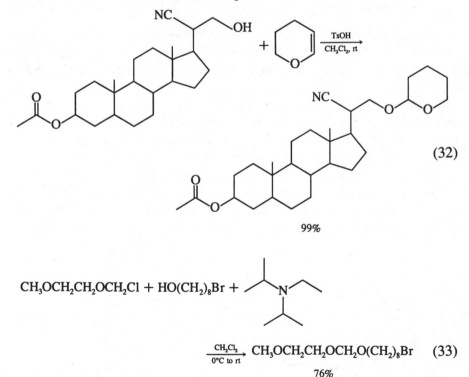

(32)

99%

$$CH_3OCH_2CH_2OCH_2Cl + HO(CH_2)_8Br +$$

$$\xrightarrow[\text{0°C to rt}]{CH_2Cl_2} CH_3OCH_2CH_2OCH_2O(CH_2)_8Br \quad (33)$$

76%

Where the acetals were used as temporary protecting groups, they may be removed with aqueous acid to recover the ketone or aldehyde. Deprotection of alcohols is done in aqueous or alcoholic acid. The meth-

oxyethoxymethyl ethers (Eq. 33) can be removed specifically by anhy
drous zinc bromide in dichloromethane at 25°C followed by aqueous
bicarbonate, conditions that leave other acetals intact.[57] This allows concurrent protection of two different alcohol functions in a molecule, and
the selective deprotection of one of them at an appropriate stage in a
synthesis.

## 4.2.5 Vinyl Ethers

Vinyl ethers, also known as *enol ethers*, are generally prepared from
acetals by acid-catalyzed elimination of an alcohol. A ketone may be
treated with methyl orthoformate in the presence of a catalytic amount
of *p*-toluenesulfonic acid to produce the acetal, and then heated directly
to cause elimination of the alcohol:[58]

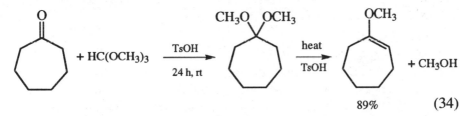

$$89\% \qquad (34)$$

The acetals may be cleaved at $-20°C$ to room temperature using trimethylsilyl trifluoromethanesulfonate and *N,N*-diisopropylethylamine,
to afford the vinyl ethers in 89–98% yields along with the alkyl trimethylsilyl ether.[59] Alkoxide catalyzed addition of alcohols to acetylene,
and reaction of alkoxides with vinyl halides also afford vinyl ethers.[60]

Silyl enol ethers are usually prepared by treating a ketone with trimethylsilyl chloride and triethylamine in refluxing DMF. In unsymmetric ketones, this gives the more substituted double bond (Eq. 35).[61]

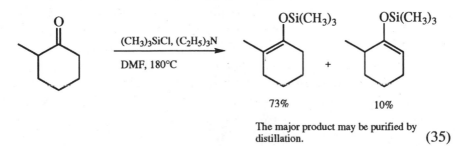

73%            10%

The major product may be purified by
distillation.                                    (35)

When the less substituted product is desired, it is made from the less
substituted enolate prepared under nonequilibrium conditions where the
less hindered proton is removed with a bulky base (Eq. 36). Selectivity

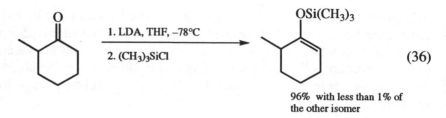

$$\text{(36)}$$

96% with less than 1% of
the other isomer

and yields may be improved by adding the trimethylchlorosilane before the LDA, at $-78°C$.[62]

These enol derivatives are used for formation of carbon–carbon bonds at the $\alpha$ position (Section 5.1.2). They also make the enol double bond available for oxidative cleavage or use in Diels–Alder reactions (Section 6.6).

## 4.3  ALCOHOLS

Reduction of acids, acid derivatives, aldehydes, and ketones gives alcohols. Esters and acids can be reduced with lithium aluminum hydride. Where selectivity is needed, carboxylic acids may be reduced with sodium borohydride and iodine even in the presence of esters (Eq. 37).[63]

$$\underset{\text{CH}_3\text{OC(CH}_2)_8\text{COH}}{\overset{\text{O}\quad\quad\text{O}}{\parallel\quad\quad\parallel}}
\quad
\begin{array}{c}\text{1. NaBH}_4\text{, THF, rt}\\[4pt]\longrightarrow\\[2pt]\text{2. I}_2\text{, THF, 0°C}\\ \text{3. aq. HCl}\end{array}
\quad
\underset{\underset{89\%}{\text{CH}_3\text{OC(CH}_2)_8\text{CH}_2\text{OH}}}{\overset{\text{O}}{\overset{\parallel}{\phantom{x}}}}
\quad\text{(37)}$$

Alkenes are unaffected. Diborane will also selectively reduce carboxylic acids in the presence of esters.[64] Aldehydes and ketones are reduced with sodium borohydride or hydrogen over platinum.

At the same oxidation level, alcohols can be prepared by substitution reactions and addition reactions. Alkali hydroxides will convert appropriate alkyl chlorides, bromides, iodides, and sulfonates to alcohols. Acid-catalyzed addition of water to alkenes gives Markovnikov alcohols, and hydroboration followed by oxidation gives anti-Markovnikov alcohols. Hindered boranes such as 9-borabicyclo[3.3.1]nonane (9-BBN) are used when selectivity toward one double bond, or higher regioselectivity is needed (Eq. 38).[65]

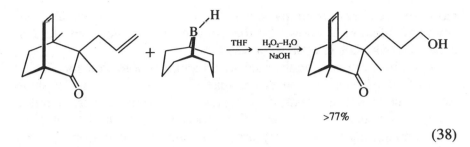

>77%

(38)

## 4.4  ETHERS

Like alcohols, ethers are commonly prepared by nucleophilic substitution or by addition to alkenes. In the Williamson method an alkoxide will displace a halide or sulfate group from a primary carbon. Suitable concentrations of alkoxides are available by using sodium hydroxide as a slurry in DMSO,[66] or in aqueous solution with a phase-transfer catalyst (Section 9.6). In the example in Eq. 39, an excess of dibromobutane was used to minimize formation of a diether.[67]

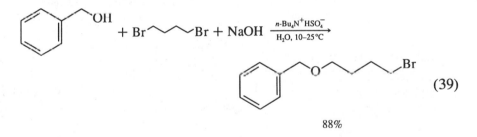

88%

(39)

Overall addition of alcohols to alkenes is accomplished by alkoxy-mercuration followed by reduction as shown in Eq. 40 for the prepa-

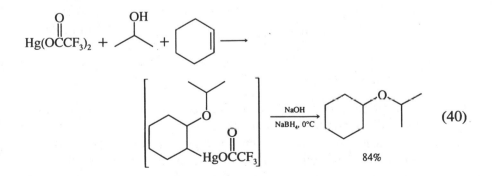

84%

(40)

ration of cyclohexyl isopropyl ether.[68] The process gives overall Markovnikov addition as shown by the conversion of 1-dodecene to 2-*n*-dodecyl ethyl ether in 80% isolated yield.

Silyl ethers are commonly used as temporary protection for alcohol groups.[69] Trimethylsilyl ethers are unstable toward mild nucleophiles such as methanol, especially in the presence of acid, but *tert*-butyldimethylsilyl ethers are stable toward a variety of mildly basic, reducing, and oxidizing conditions. They are commonly prepared from primary and secondary alcohols by treatment with *tert*-butyldimethylchlorosilane and imidazole in THF. The chlorosilane and a catalytic amount of 4-dimethylaminopyridine can be used to selectively silylate primary alcohols in the presence of secondary alcohols (Eq. 41).[70] In this example,

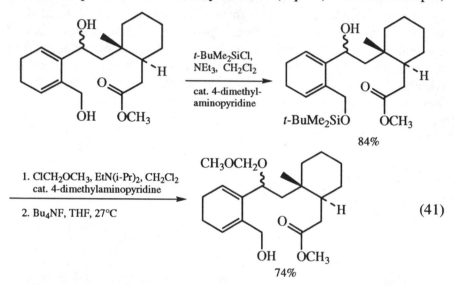

(41)

the protection at the primary alcohol was used temporarily to allow acetal protection at the secondary alcohol. The silyl protecting group may be removed to free the alcohol by using tetrabutylammonium fluoride in THF (Eq. 41), taking advantage of the high affinity of fluoride for silicon. Dilute HF or acetic acid serve also.

## 4.5   ALKYL HALIDES

Chlorine or bromine may be incorporated by substitution for hydrogen under free-radical conditions. This is useful where the substrate is a highly symmetric compound or contains a site especially prone to free-radical formation. Otherwise complex mixtures of isomers are obtained.

Anhydrous hydrogen chloride, bromide, or iodide will add to alkenes by a carbocationic mechanism to give Markovnikov products. The products may have rearranged structures if the first intermediate carbocation can improve in stability by a 1,2-hydride or alkyl shift. Addition to $\alpha,\beta$-unsaturated carbonyl compounds affords the $\beta$-halo compounds (Eq. 42).[71] Aqueous hydrogen halides may be used with alkenes by heating

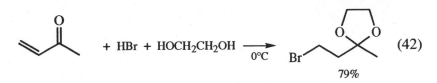

in the presence of a phase-transfer catalyst such as hexadecyltributyl-phosphonium bromide (Eq. 43).[72] Only Markovnikov products are formed, even with hydrobromic acid and added peroxides.

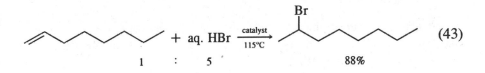

The most versatile and specific routes begin with alcohols. These and some other methods are given in the following sections.

## 4.5.1  Alkyl Chlorides

Iminium salts (Vilsmeier reagents), prepared from inorganic acid chlorides and amides give high yields of alkyl chlorides from primary and secondary alcohols. The salts may be isolated as crystalline solids (Eq. 44)[73] and then combined with the alcohols, or they may be prepared *in situ* (Eq. 45).[74] The process occurs with close to 100% inversion of configuration and without rearrangement.

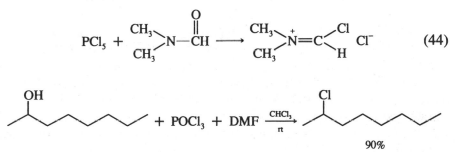

A very mild, neutral procedure for 1° and 2° alcohols is treatment with triphenylphosphine and carbon tetrachloride (Eq. 46).[75] In pyridine this is selective for 1° in the presence of 2° alcohol functions. Rearrangement is absent even in the conversion of neopentyl alcohol to chloroneopentane.

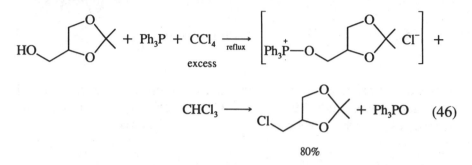

80%

Some tertiary alcohols may be converted to the chlorides by simply shaking with concentrated aqueous HCl. This involves carbocations and may lead to rearrangements.

## 4.5.2  Alkyl Bromides

As with chlorides, the iminium salts are particularly useful for converting primary and secondary alcohols to bromides.[76] Phosphorus tribromide and DMF give the analogous reagent which, in dioxane, converted $(-)$-2-octanol to $(+)$-2-bromooctane in 85% yield with some loss of optical purity but no rearrangement.[73]

Triphenylphosphine with carbon tetrabromide at ice-bath temperatures gives good yields of primary alkyl bromides but is poor with most secondary alcohols.[77]

Primary, secondary, and tertiary alcohols are converted in high yield to the bromides by treatment with trimethylsilyl bromide in chloroform at 25–50°C.[78] Again the stereochemistry is mostly inversion (93.8% with 2-octanol).

Allylic hydrogens may be replaced with bromine by using N-bromosuccinimide under sunlamp irradiation. This is a free-radical chain mechanism, and two structural isomers may result from attachment at either end of the conjugated radical system. Benzylic hydrogens may be replaced similarly with N-bromosuccinimide, or simply with $Br_2$ in carbon tetrachloride in the presence of a catalytic amount of solid $La(OCOCH_3)_3$. With this catalyst, room fluorescent lighting is sufficient to initiate the reaction (Eq. 47). Yields are 50–90%.[79]

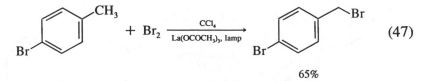

(47)

65%

Ester enolate anions may be treated with excess 1,2-dibromoethane to afford the α-bromoesters.[80] Ketones may be α-brominated via the

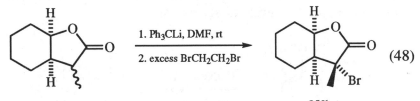

(48)

85%

silyl enol ether.[81]

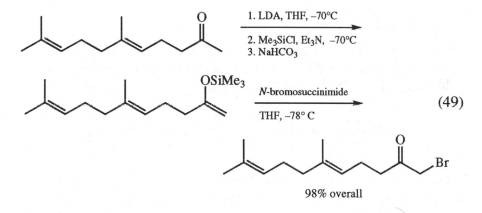

(49)

98% overall

### 4.5.3  Alkyl Iodides

Primary, secondary, and tertiary alkyl iodides may be prepared from alcohols by treatment with a solution of $P_2I_4$ in carbon disulfide at 0°C in yields of 45–90%, again without rearrangement.[82] Primary and secondary alcohols may be treated with a solution of triphenylphosphine, imidazole, and iodine in dichloromethane at room temperature to give high yields of alkyl iodides with about 80% inversion in secondary cases.[83]

The classic Finkelstein synthesis of primary iodides begins with the corresponding chloride, bromide, or tosylate. Treatment with a dry acetone solution of sodium iodide gives a precipitate of the corresponding sodium salt plus the alkyl iodide (Eq. 50).[84]

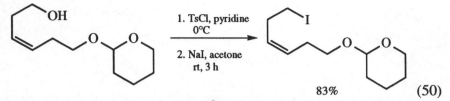

$$(50)$$

83%

Ketone enol silyl ethers or acetates are readily converted to α-iodoketones in high yield by treatment with a molar equivalent if iodine and copper(II) nitrate in acetonitrile.[85]

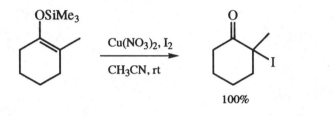

100%

$$(51)$$

### 4.5.4 Alkyl Fluorides

Primary, secondary, and tertiary alcohols can be converted directly to fluorides, usually in high yield, by treatment with diethylaminosulfur trifluoride[86] (Eq. 52).[84] Most cases proceed with inversion of configu-

$$Br(CH_2)_8OH + (C_2H_5)_2NSF_3 \xrightarrow[-78° \text{ to } 25°C]{CH_2Cl_2} Br(CH_2)_8F$$

61%

$$(52)$$

ration and without rearrangement. Retention may be found when a chlorine or bromine is present on an adjacent carbon. Aldehydes and ketones will also react under somewhat more vigorous conditions to afford geminal difluoro compounds and some vinyl fluorides. Selectivity is such that keto alcohols may be converted to keto fluoro compounds. The reagent is prepared by combining diethylaminotrimethylsilane and SF$_4$ at low temperature, and it is also commercially available. It can decompose violently above 50°C.

Primary and secondary chlorides, bromides, and tosylates may be converted to fluorides in isolated yields of 35–66% by stirring with almost anhydrous tetrabutylammonium fluoride at 25–40°C without solvent.[87] The by-products include alcohols from the small amount of water present, and alkenes arising because of the high basicity of unsolvated fluoride.

Lithium enolates from esters, amides, and ketones may be fluorinated with N-fluorobis[(trifluoromethyl)sulfonyl]imide.[88]

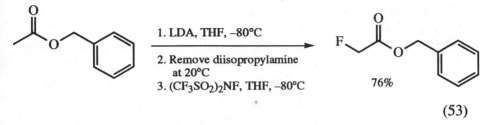

$$\text{1. LDA, THF, } -80°C$$
$$\text{2. Remove diisopropylamine at 20°C}$$
$$\text{3. (CF}_3\text{SO}_2)_2\text{NF, THF, } -80°C$$

76%

(53)

## 4.6 AMINES

Amines are commonly prepared by treating primary or secondary halides or sulfonates with excess ammonia. The primary amine produced may compete with ammonia to give some secondary and even tertiary amine. The excess is used to minimize this. A secondary or tertiary amine may be prepared by treating a primary or secondary amine with an alkyl halide or sulfonate in the same way. Primary amines can be made cleanly using potassium phthalimide instead of ammonia (Chapter 9, Eq. 10). The free amine may be released from the alkylated phthalimide by shaking briefly with aqueous methylamine at room temperature.[89] This gives N,N'-dimethylphthalamide and the desired amine.

Reductive amination of ketones and aldehydes gives primary, secondary, and tertiary amines. Sodium cyanoborohydride[90] or platinum-catalyzed hydrogenation are used as shown in Eqs. 54[91] and 55.[92]

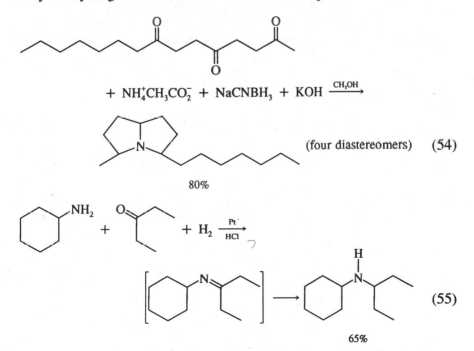

$$+ \text{ NH}_4^+\text{CH}_3\text{CO}_2^- + \text{ NaCNBH}_3 + \text{ KOH} \xrightarrow{\text{CH}_3\text{OH}}$$

(four diastereomers) (54)

80%

65%

(55)

Tertiary alkyl groups cannot be attached to nitrogen by the preceding reactions, and one must resort to carbocation methods. In the Ritter reaction an alcohol is treated with a strong acid in the presence of a nitrile. The alcohol is converted to a carbocation that attacks the nitrogen of the nitrile to give, after hydration, a hydrolyzable amide (Eq. 56).[93]

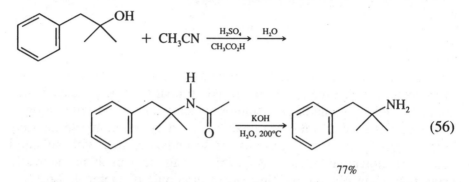

(56)

77%

Other nitrogen-containing functional groups may be reduced to give amines. Nitriles may be hydrogenated over a platinum or palladium catalyst at 25°C. This process can give some secondary amine via addition of primary amine to the intermediate imine, but that reaction can be suppressed by providing an acid to render the salt of the primary amine. Oximes may also be hydrogenated to primary amines over rhodium on alumina. Lithium aluminum hydride or Na-AlH$_2$(OCH$_2$CH$_2$OCH$_3$)$_2$ will reduce amides to secondary or tertiary amines and will also reduce nitriles, oximes, and nitroalkanes to primary amines.

## 4.7  ISOCYANATES

Isocyanates are commonly prepared by treating primary amine hydrochlorides with phosgene in a hot solvent. Gaseous hydrogen chloride is eliminated from the intermediate carbamoyl chlorides. On a laboratory scale it is more convenient to use bis-trichloromethyl carbonate, "triphosgene," a stable, easily weighed solid, mp 81–83°C (Eq. 57).[94] One-third of a mole of triphosgene serves for one mole of phosgene.

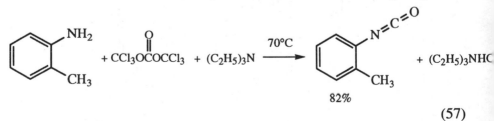

82%

(57)

Carboxylic acid chlorides may be converted to isocyanates via the Curtius rearrangement of the acyl azide. A solution of tetrabutylammonium azide (prepared by $CH_2Cl_2$ extraction from aqueous sodium azide and tetrabutylammonium hydroxide) in toluene or benzene reacts readily with acid chlorides to give acyl azides. These are heated in solution to afford the isocyanates and nitrogen (Eq. 58).[95]

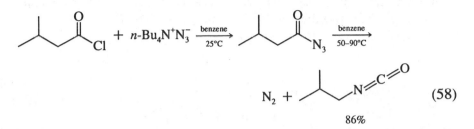

$$N_2 + \qquad\qquad\qquad\qquad\qquad (58)$$

86%

## 4.8 ALKENES

Alkenes may be made from saturated compounds by various $\beta$-elimination reactions. A vicinal dihalide may be dehalogenated by sodium iodide or activated zinc to give a double bond specifically between the carbons that bore the halogens. One practical use of such a process is the inversion of configuration of alkenes. The anti addition of chlorine followed by net syn elimination using sodium iodide does so as in Eq. 59.[96] About 90% anti elimination may be obtained with activated zinc and acetic acid in DMF.

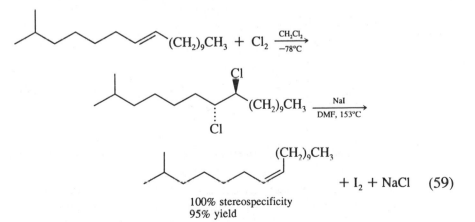

$$+ I_2 + NaCl \quad (59)$$

100% stereospecificity
95% yield

The more common circumstance is the elimination of a leaving group and a hydrogen. $\beta$-Dehydrohalogenation is usually done with strong bases. If substitution competes, sterically hindered, less nucleophilic

bases such as potassium *tert*-butoxide are chosen. Here regiospecificity
is a problem because there is usually a choice of $\beta$ hydrogens. Anti
elimination is usual where conformationally allowed. Among anti possibilities, conjugated products predominate over nonconjugated, and
more substituted alkenes predominate over less substituted ones. If the
alkene will be conjugated, less potent bases may be used, especially
when there are other base-sensitive functional groups in the molecule.
The amidine 1,5-diazabicyclo[4.3.0]non-5-ene (DBN)[97] or LiBr and
Li$_2$CO$_3$ in DMF (Eq. 60)[98] give good selectivity.

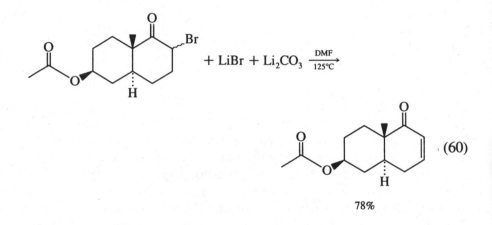

$$78\%$$

Alcohols may be eliminated by acid treatment. Rearrangements are
likely because carbocations are intermediates (Eq. 61).[99]

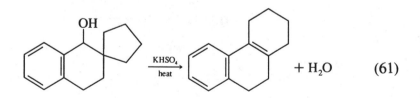

Selenoxides eliminate readily without a base. They are generally prepared from enolate anions by reaction with diphenyldiselenide or phen-
ylselenyl bromide to give phenylselenides. The phenylselenides are oxidized with sodium periodate, hydrogen peroxide, or peracids to the
selenoxides, which eliminate even at room temperature to afford $\alpha,\beta$-
unsaturated ketones and esters[100] (Eq. 62).[101]

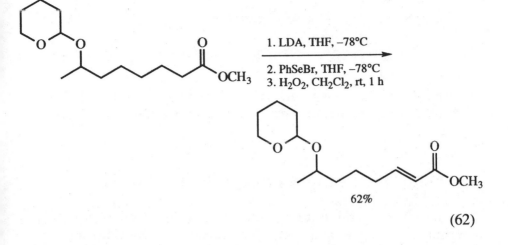

$$(62)$$

Ketones may be reductively eliminated via their tosylhydrazones[102] as illustrated in Eq. 63.[103] If there is a choice, the least substituted alkene

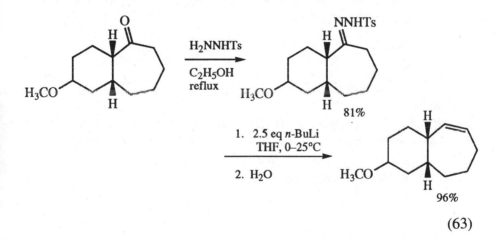

$$(63)$$

will predominate. Trisubstituted alkenes require LDA in place of the alkyllithium.[104]

 cis-Alkenes may be prepared by partial hydrogenation of appropriate alkynes. Various catalysts have been used, including colloidal nickel containing some boron from $NaBH_4$ reduction of nickel acetate. Combined with ethylene diamine, this catalyst gives a 200 : 1 selectivity toward cis isomers (Eq. 64).[105] The corresponding trans isomers are

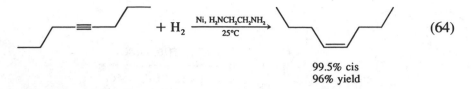

$$(64)$$

99.5% cis
96% yield

available from reduction of alkynes with sodium in liquid ammonia. Carbanions will also add to certain alkynes to give alkenes (Chapter 5, Eq. 12).

## 4.9  REDUCTIVE REMOVAL OF FUNCTIONALITY

It is sometimes necessary to reduce a functionalized site to a $CH_3$ or $CH_2$ group. Starting from the acid level of oxidation, this is usually done in two stages. For example, an ester may be reduced to an alcohol with $LiAlH_4$, the alcohol converted to a tosylate, and then reduced again to a $CH_3$ group. Ketones and aldehydes can be reduced completely via the tosylhydrazone using catecholborane followed by decomposition of the hydroboration intermediate:[106]

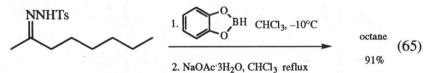

$$(65)$$

Sodium cyanoborohydride in DMF–sulfolane will reduce ketone and aldehyde tosylhydrazones selectively in the presence of esters and nitriles. One may even begin with the carbonyl compound and prepare the hydrazone in the same reaction pot:[107]

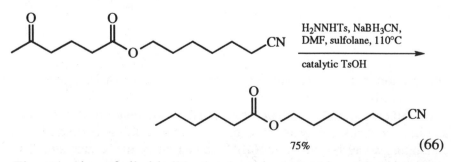

75%                    $$(66)$$

The reduction of alkyl halides has been important in many syntheses. Sodium cyanoborohydride in hexamethylphosphoric triamide will reduce alkyl iodides, bromides, and tosylates selectively in the presence of ester, amide, nitro, chloro, cyano, alkene, epoxide, and aldehyde groups.[108] Tri-*n*-butyltin hydride will replace Cl, Br, or I with hydrogen via a free-radical chain mechanism initiated by thermal decomposition of AIBN.[109] Other functionality such as ketones, esters, amides, ethers, and alcohols survive unchanged. For example, see Chapter 6, Eq. 24. Tris(trimethylsilyl) silane can be used similarly.[110]

## PROBLEMS

Show how you would prepare each of the following products from the given starting materials. Where more than one step is required, show each step distinctly.

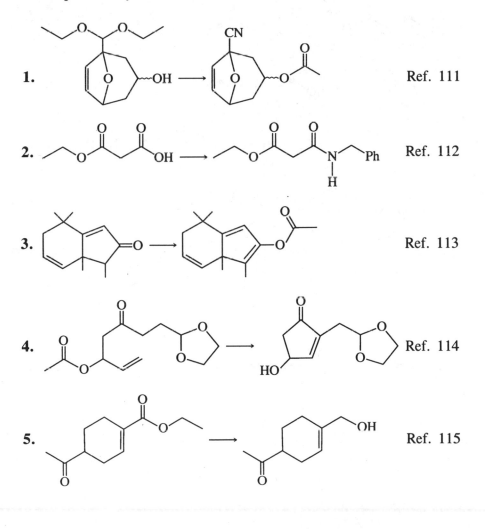

1.    Ref. 111

2.    Ref. 112

3.    Ref. 113

4.    Ref. 114

5.    Ref. 115

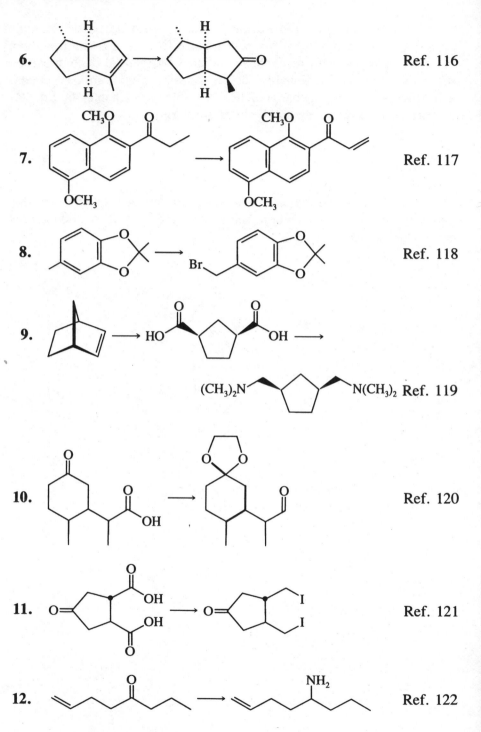

6.         Ref. 116

7.         Ref. 117

8.         Ref. 118

9.         Ref. 119

10.         Ref. 120

11.         Ref. 121

12.         Ref. 122

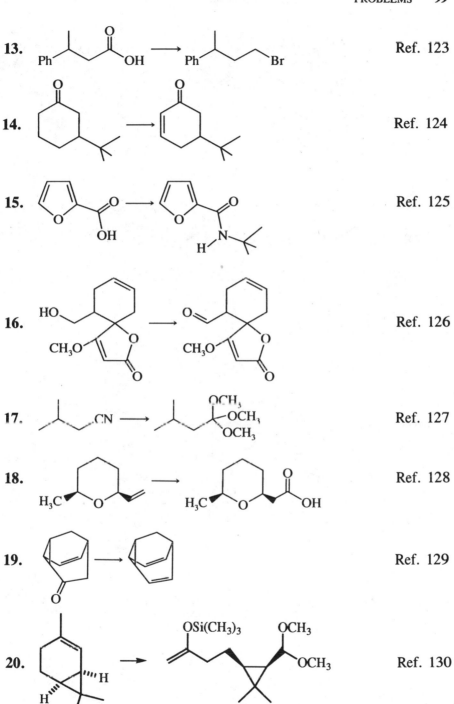

13.   Ref. 123

14.   Ref. 124

15.   Ref. 125

16.   Ref. 126

17.   Ref. 127

18.   Ref. 128

19.   Ref. 129

20.   Ref. 130

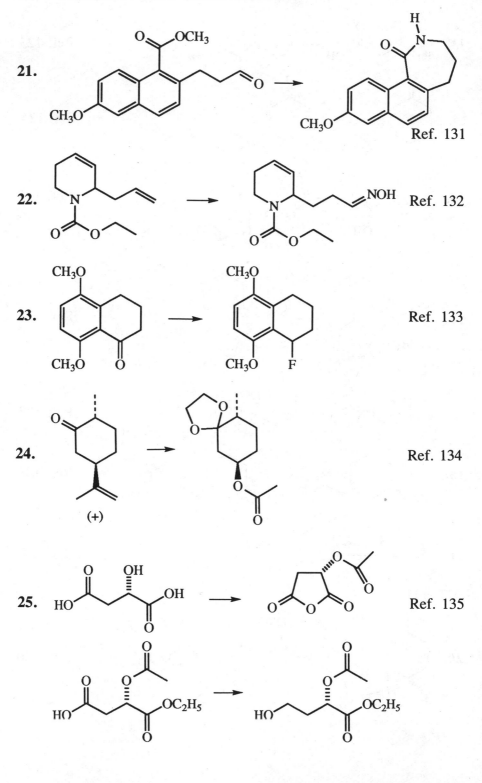

21.     Ref. 131

22.     Ref. 132

23.     Ref. 133

24.     (+)     Ref. 134

25.     Ref. 135

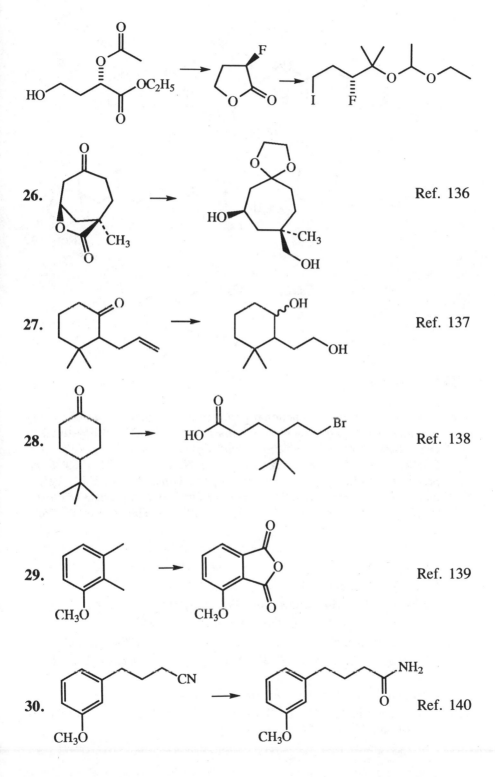

**26.**                                              Ref. 136

**27.**                                              Ref. 137

**28.**                                              Ref. 138

**29.**                                              Ref. 139

**30.**                                              Ref. 140

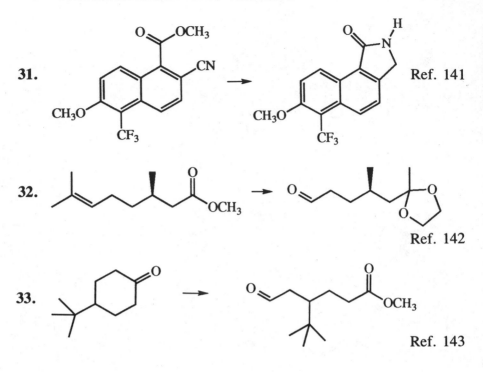

31.    Ref. 141

32.    Ref. 142

33.    Ref. 143

## REFERENCES

1. General references on functional group synthesis include *The Chemistry of Functional Groups*, multiple volumes, Patai, S., Ed., Wiley-Interscience, New York; *Organic Functional Group Preparations*, Vols. 1–3, Sandler, S. R.; Karo, W., Academic Press, New York, 1971–1983; *Comprehensive Organic Chemistry*, Barton, D.; Ollis, W. D., Eds., Pergamon Press, Oxford, 1979; *Methoden der Organischen Chemie (Houben-Weyl)*, multiple volumes, Müller, E., Ed., Georg Thieme Verlag, Stuttgart.
2. Ferguson, L.; Wims, A. I. *J. Org. Chem.* **1960**, *25*, 668.
3. Treley, W. F.; Marvel, C. S. *Org. Synth.* **1955**, Coll. *3*, 822.
4. Hertz, W.; Mohanraj, S. *J. Org. Chem.* **1980**, *45*, 5417.
5. Starks, C. M. *J. Am. Chem. Soc.* **1971**, *93*, 195.
6. Corey, E. J.; Schmidt, G. *Tetrahedron Lett.* **1979**, 399.
7. Matsuda, I.; Murata, S.; Izumi, Y. *J. Org. Chem.* **1980**, *45*, 237.
8. Nicolaou, K. C.; Pavia, M. R.; Seitz, S. P. *Tetrahedron Lett.* **1979**, 2327.
9. Overman, L. E. *Tetrahedron Lett.* **1975**, 1149.
10. Hase, T. A.; Nylund, E.-L. *Tetrahedron Lett.* **1979**, 2633; Lindgren, B. O.; Nilsson, T. *Acta Chem. Scand.* **1973**, *27*, 888.

11. Nugent, W. A.; McKinney, R. J. *J. Org. Chem.* **1985,** *50,* 5370.

12. Degenhardt, C. R. *J. Org. Chem.* **1980,** *45,* 2763.

13. Burgstahler, A. W.; Weigel, L. O.; Bell, W. J.; Rust, M. K. *J. Org. Chem.* **1975,** *40,* 3456.

14. Jones, G.; Raphael, R. A.; Wright, S. *J. Chem. Soc. Perkin I* **1974,** 1676.

15. Mehta, G.; Pandey, P. N. *Synthesis* **1975,** 404.

16. House, H. O.; Gall, M.; Olmstead, H. D. *J. Org. Chem.* **1971,** *36,* 2361.

17. Guthrie, J. L.; Rabjohn, N. *Org. Synth.* **1963,** *Coll 4,* 513.

18. Inami, K.; Shiba, T. *Bull. Chem. Soc. Jpn.,* **1985,** *58,* 352.

19. Cacchi, S.; Misiti, D. *Synthesis* **1980,** 243.

20. Ravindranathan, M.; Kalyanam, N.; Sivaram, S. *J. Org. Chem.* **1982,** *47,* 4812.

21. Adams, R.; Ulich, L. H. *J. Am. Chem. Soc.* **1920,** *42,* 599.

22. Burgstahler, A. W.; Weigel, L. O.; Shaefer, C. G. *Synthesis* **1976,** 767.

23. Hurd, C. D.; Christ, R.; Thomas, C. L. *J. Am. Chem. Soc.* **1933,** *55,* 2589.

24. Krimen, L. I. *Org. Synth.* **1970,** *50,* 1.

25. Fife, W. K.; Zhang, Z. *J. Org. Chem.* **1986,** *51,* 3744.

26. Bargar, T. M.; Riley, C. M. *Synth. Commun.* **1980,** *10,* 479.

27. Olah, G. A.; Narang, S. C.; Fung, A. P.; Gupta, B. G. B. *Synthesis* **1980,** 657.

28. Olah, G. A.; Vankar, Y. D.; Garcia-Luna, A. *Synthesis* **1979,** 227.

29. Ho, T. L.; Wong, C. M. *Synth. Commun.* **1975,** *5,* 299.

30. DeWolfe, R. H. *Carboxylic Ortho Acid Derivatives,* Academic Press, New York, **1970;** DeWolfe, R. H. *Synthesis* **1974,** 153.

31. McElvain, S. M.; Nelson, J. W. *J. Am. Chem. Soc.* **1942,** *64,* 1825.

32. McElvain, S. M.; Venerable, J. T. *J. Am. Chem. Soc.* **1950,** *72,* 1661.

33. Corey, E. J.; Raju, N. *Tetrahedron Lett.* **1983,** *24,* 5571.

34. Burgstahler, A. W.; Weigel, L. O.; Shaefer, C. G. *Synthesis* **1986,** 767.

35. Johnstone, R. A. W.; Telford, R. P. *J. Chem. Soc. Chem. Commun.* **1978,** 354.

36. Marshall, J. A.; Crooks, S. L.; DeHoff, B. S. *J. Org. Chem.* **1988,** *53,* 1616.

37. Ireland, R. E.; Courtney, L.; Fitzsimmons, B. J. *J. Org. Chem.* **1983,** *48,* 5186.

38. Malek, J.; Cerny, M. *Synthesis* **1972,** 217.

39. Corey, E. J.; Suggs, J. W. *Tetrahedron Lett.* **1975,** 2647.

40. Tidwell, T. T. *Org. Reactions* **1990,** *39,* Chapter 3.

41. Hecker, S. J.; Heathcock, C. H. *J. Org. Chem.* **1985,** *50,* 5159.

42. Guziec, T. S., Jr.; Luzzio, F. A. *J. Org. Chem.* **1982,** *47,* 1787.

43. Pappas, J. J.; Keaveney, W. P.; Gancher, E.; Berger, M. *Tetrahedron Lett.* **1966,** 4273.

44. Maas, D. D.; Blagg, M.; Wiemer, D. F. *J. Org. Chem.* **1984,** *49,* 853.

45. Wiberg, K. B. in *Oxidation in Organic Chemistry, Part A*, Wiberg, K. B., Ed., Academic Press, New York, **1965,** Chapter 2.

46. Stevens, R. V.; Chapman, K. T.; Weller, H. N. *J. Org. Chem.* **1980,** *45,* 2030.

47. Mohrig, J. R.; Nienhuis, D. M.; Linch, C. F.; Van Zoeren, C.; Fox, B. G.; Mahaffy, P. G. *J. Chem. Ed.* **1985,** *62,* 519.

48. Yoshioka, H.; Takasaki, K.; Kobayashi, M.; Matsumoto, T. *Tetrahedron Lett.* **1979,** 3489; Martin, S. F.; Chou, T. *Tetrahedron Lett.* **1978,** 1943; Julia, M.; Blasioli, C. *Bull. Soc. Chim. Fr.* **1976,** 1941.

49. Layer, R. W. *Chem. Rev.* **1963,** *63,* 489.

50. Pearce, G. T.; Gore, W. E.; Silverstein, R. M. *J. Org. Chem.* **1976,** *41,* 2797.

51. Hünig, S.; Lüche, E.; Brenninger, W. *Org. Synth.* **1973,** *Coll 5,* 808.

52. Whitesell, J. K.; Whitesell, M. A. *Synthesis* **1983,** 517; *Enamines*, 2nd ed., Cook, A. G., Ed., Marcell Dekker, New York, 1988.

53. Meskens, F. A. J. *Synthesis* **1981,** 501.

54. Lansbury, P. T.; Mazur, D. J. *J. Org. Chem.* **1985,** *50,* 1632.

55. Bauduin, G.; Pietrasanta, Y. *Tetrahedron* **1973,** *29,* 4225.

56. Millar, J. G.; Oehlschlager, A. C.; Wong, J. W. *J. Org. Chem.* **1983,** *48,* 4404.

57. Corey, E. J.; Gras, J. L.; Ulrich, P. *Tetrahedron Lett.* **1976,** 809.

58. Wohl, R. A. *Synthesis,* **1974,** 38.

59. Gassman, P. G.; Burns, S. J. *J. Org. Chem.* **1988,** *53,* 5574.

60. Meerwein, H. in Houben-Weyl, *Methoden der Organischen Chemie*, Vol. 6/3, Müller, E., Ed., Georg Thieme Verlag, Stuttgart, 1965, pp. 90–116.

61. House, H. O.; Czuba, L. J.; Gall, M.; Olmstead, H. D. *J. Org. Chem.* **1969,** *34,* 2324; Paterson, I.; Fleming, I. *Tetrahedron Lett.* **1979,** 995.

62. Corey, E. J.; Gros, A. W. *Tetrahedron Lett.* **1984,** *25,* 495.

63. Kanth, J. V. B.; Periasamy, M. *J. Org. Chem.* **1991,** *56,* 5964.

64. Frye, S. V.; Eliel, E. L. *J. Org. Chem.* **1985,** *50,* 3402.

65. Snowden, R.; Sonnay, P. *J. Org. Chem.* **1984,** *49,* 1464.

66. Benedict, D. R.; Bianchi, T. A.; Cate, L. A. *Synthesis* **1979,** 428.

67. Burgstahler, A. W.; Weigel, L. O.; Sanders, M. E.; Shaefer, C. G.; Bell, W. J.; Vuturo, S. B. *J. Org. Chem.* **1977,** *42,* 566.

68. Brown, H. C.; Kurek, J. T.; Rei, M.-H.; Thompson, K. L. *J. Org. Chem.* **1985,** *50,* 1171.

69. Lalonde, M.; Chan, T. H. *Synthesis* **1985,** 817.

70. Funk, R. L.; Daily, W. J.; Parvez, M. *J. Org. Chem.* **1988,** *53*, 4141.

71. Rigby, J. H.; Wilson, J. A. Z. *J. Org. Chem.* **1987,** *52*, 34.

72. Landini, D.; Rolla, F. *J. Org. Chem.* **1980,** *45*, 3527.

73. Hepburn, D. R.; Hudson, H. R. *J. Chem. Soc. Perkin I* **1976,** 754.

74. Yoshihara, M.; Eda, T.; Sakaki, K.; Maeshima, F. *Synthesis* **1980,** 746.

75. Castro, B. R. *Org. Reactions* **1983,** *29*, 1; Lee, J. B.; Nolan, T. J. *Can. J. Chem.* **1966,** *44*, 1331.

76. Dolbier, Jr., W. R.; Dulcere, J.-P.; Sellers, S. F.; Korniak, H.; Shatkin, B. T.; Clark, T. L. *J. Org. Chem.* **1982,** *47*, 2298.

77. Kocienski, P. J.; Cernigliaro, G.; Feldstein, G. *J. Org. Chem.* **1977,** *42*, 353.

78. Jung, M. E.; Hatfield, G. L. *Tetrahedron Lett.* **1978,** 4483.

79. Ouertani, M.; Girard, P.; Kagan, H. B. *Bull. Soc. Chim. Fr.* **1982,** II-327.

80. Greene, A. E.; Muller, J.-C.; Ourisson, G. *J. Org. Chem.* **1974,** *39*, 186.

81. Hajos, Z. G.; Wachter, M. P.; Werblood, H. M.; Adams, R. E. *J. Org. Chem.* **1984,** *49*, 2600.

82. Lauwers, M.; Regnier, B.; Van Enoo, M.; Denis, J. N.; Krief, A. *Tetrahedron Lett.* **1979,** 1801.

83. Lange, G. L.; Gottardo, C. *Synth. Commun.* **1990,** *20*, 1473.

84. Carvalho, J. F.; Prestwich, G. D. *J. Org. Chem.* **1984,** *49*, 1251.

85. Dalla Cort, A. *J. Org. Chem.* **1991,** *56*, 6708.

86. Hudlicky, M. *Org. Reactions* **1988,** *35*, 513.

87. Cox, D. P.; Terpinski, J.; Lawrynowicz, W. *J. Org. Chem.* **1984,** *49*, 3216.

88. Resnati, G.; DesMarteau, D. D. *J. Org. Chem.* **1991,** *56*, 4925.

89. Wolfe, S.; Hasan, S. K. *Can. J. Chem.* **1970,** *48*, 3572.

90. Lane, C. F. *Synthesis* **1975,** 135.

91. Jones, T. H.; Blum, M. S.; Fales, H. M.; Thompson, C. R. *J. Org. Chem.* **1980,** *45*, 4778.

92. Stowell, J. C.; Padegimas, S. J. *J. Org. Chem.* **1974,** *39*, 2448.

93. Timberlake, J. W.; Alender, J.; Garner, A. W.; Hodges, M. L.; Ozmeral, C.; Szilagyi, S.; Jacobus, J. O. *J. Org. Chem.* **1981,** *46*, 2082.

94. Eckert, H.; Forster, B. *Angew. Chem. Internatl. Ed.* **1987,** *26*, 894.

95. Brändström, A.; Lamm, B.; Palmertz, I. *Acta Chem. Scand. B* **1974,** *28*, 699.

96. Sonnet, P. E.; Oliver, J. E. *J. Org. Chem.* **1976,** *41*, 3284, 3279.

97. Oediger, H.; Moller, F.; Eiter, K. *Synthesis* **1972,** 591.

98. Kametani, T.; Suzuki, K.; Nemoto, H. *J. Org. Chem.* **1980,** *45*, 2204.

99. Christol, H.; Jacquier, R.; Mousseron, M. *Bull. Soc. Chim. Fr.* **1958,** 248.

100. Liotta, D. *Acc. Chem. Res.* **1984,** *17,* 28.

101. Ngooi, T. K.; Scilimati, A.; Guo, Z.; Sih, C. J. *J. Org. Chem.* **1989,** *54,* 911.

102. Shapiro, R. H. *Org. React.* **1976,** *23,* 405.

103. Gleiter, R.; Müller, G. *J. Org. Chem.* **1988,** *53,* 3912.

104. Kolonko, K. J.; Shapiro, R. H. *J. Org. Chem.* **1978,** *43,* 1404.

105. Brown, C. A.; Ahuja, V. K. *J. Chem. Soc. Chem. Commun.* **1973,** 553.

106. Kabalka, G. W.; Baker, Jr., J. D. *J. Org. Chem.* **1975,** *40,* 1834.

107. Hutchins, R. D.; Milewski, C. A.; Maryanoff, B. E. *J. Am. Chem. Soc.* **1973,** *95,* 3662.

108. Lane, C. F. *Synthesis* **1975,** 135.

109. Neumann, W. P. *Synthesis* **1987,** 665–683.

110. Giese, B.; Kopping, B. *Tetrahedron Lett.* **1989,** *30,* 681.

111. La Belle, B. E.; Knudsen, M. J.; Olmstead, M. M.; Hope, H.; Yanuck, M. D.; Schore, N. E. *J. Org. Chem.* **1985,** *50,* 5215.

112. Moyer, M. P.; Feldman, P. L.; Rapoport, H. *J. Org. Chem.* **1985,** *50,* 5226.

113. Ishihara, M.; Tsuneya, T.; Shiota, H.; Shiga, M.; Nakatasu, K. *J. Org. Chem.* **1986,** *51,* 491.

114. Ellison, R. A.; Lukenbach, E. R.; Chiu, C. *Tetrahedron Lett.* **1975,** 499.

115. Mandai, T.; Yokoyama, H.; Miki, T.; Fukuda, H.; Kolata, H.; Kawada, M.; Otera, J. *Chem. Lett.* **1980,** 1057.

116. Matthews, R. S.; Whitesell, J. K. *J. Org. Chem.* **1975,** *40,* 3312.

117. Sih, C. J.; Massuda, D.; Corey, P.; Gleim, R. D.; Suzuki, F. *Tetrahedron Lett.* **1979,** 1285.

118. Ogura, K.; Tsuchihashi, G. *Tetrahedron Lett.* **1971,** 3151.

119. Wiberg, K. B.; Bailey, W. F.; Jason, M. E. *J. Org. Chem.* **1976,** *41,* 2711.

120. Ficini, J.; Eman, A.; Touzin, A. M. *Tetrahedron Lett.* **1976,** 679.

121. Baraldi, P. G.; Pollini, G. P.; Simoni, D.; Barco, A.; Benetti, S. *Tetrahedron* **1984,** *40,* 761.

122. Harding, K. E.; Burks, S. R. *J. Org. Chem.* **1984,** *49,* 40.

123. Kogura, T.; Eliel, E. L. *J. Org. Chem.* **1984,** *49,* 577.

124. Gorthey, L. A.; Vairamani, M.; Djerassi, C. *J. Org. Chem.* **1985,** *50,* 4173.

125. Carpenter, A. J.; Chadwick, D. J. *J. Org. Chem.* **1985,** *50,* 4362.

126. Takeda, K.; Shibata, Y.; Sagawa, Y.; Urahata, M.; Funaki, K.; Hori, K.; Sasahara, H.; Yoshii, E. *J. Org. Chem.* **1985,** *50,* 4673.

127. Wiberg, K.; Martin, E. J.; Squires, R. R. *J. Org. Chem.* **1985**, *50*, 4717.

128. Kim, Y.; Mundy, B. P. *J. Org. Chem.* **1982**, *47*, 3556.

129. Henkel, J. G.; Hane, J. T. *J. Org. Chem.* **1983**, *48*, 3858.

130. Taylor, M. D.; Minaskanian, G.; Winzenberg, K. N.; Santone, P.; Smith, A. B. III *J. Org. Chem.* **1982**, *47*, 3960.

131. Wrobel, J.; Dietrich, A.; Gorham, B. J.; Sestanj, K. *J. Org. Chem.* **1990**, *55*, 2694.

132. Kozikowski, A. P.; Park, P. *J. Org. Chem.* **1990**, *55*, 4668.

133. Chen, C.-P.; Swenton, J. S. *J. Org. Chem.* **1985**, *50*, 4569.

134. Solladié, G.; Hutt, J. *J. Org. Chem.* **1987**, *52*, 3560.

135. Shiuey, S.-J.; Partridge, J. J.; Uskokovic, M. R. *J. Org. Chem.* **1988**, *53*, 1040.

136. Baldwin, J. E.; Broline, B. M. *J. Org. Chem.* **1982**, *47*, 1385.

137. Magatti, C. V.; Kaminski, J. J.; Rothberg, I. *J. Org. Chem.* **1991**, *56*, 3102.

138. Paquette, L. A.; Ra, C. S. *J. Org. Chem.* **1988**, *53*, 4978.

139. Hansen, D. W., Jr.; Pappo, R.; Garland, R. B. *J. Org. Chem.* **1988**, *53*, 4244.

140. Venit, J. J.; DiPerro, M.; Magnus, P. *J. Org. Chem.* **1989**, *54*, 4298.

141. Wrobel, J.; Dietrich, A.; Gorham, B. J.; Sestanj, K. *J. Org. Chem.* **1990**, *55*, 2694.

142. Chiang, Y.-C. P.; Yang, S. S.; Heck, J. V.; Chabala, J. C.; Chang, M. N. *J. Org. Chem.* **1989**, *54*, 5708.

143. Beckwith, A. L. J.; Zimmermann, J. *J. Org. Chem.* **1991**, *56*, 5791.

# 5

# CARBON–CARBON BOND FORMATION

For two carbons to be mutually attractive and join together, they usually begin with opposite charge polarizations. One has available electrons and is termed the *nucleophile* (seeking a plus charge or nucleus), and the other carries a partial or full positive charge and is called the *electrophile*. In basic solution the nucleophile carries a negative charge because it is bonded to, or associated ionically with, a more electropositive element, that is, a metal. This connection may be direct or may include intervening $\pi$-conjugated atoms. The electrophile has a partially positive carbon because of a dipolar bond with a more electronegative atom. In acid solution the nucleophile is an alkene or an arene with its projecting $\pi$ electrons, with or without polarization (but not a carbanion because acid solutions simply protonate them). In acid the electrophile is a complexed or free carbocation. Examples of each are shown in Table I.

Some carbon–carbon bond-forming reactions occur with little or no charge polarization. Examples include the Diels–Alder, Cope, and Claisen reactions where concerted bond reorganization occurs (Chapter 8). Others involve the transient neutral but electrophilic intermediates, carbenes, and arynes.

Example applications of these in the formation of single and double bonds are shown in the following sections and in Chapter 6.[1,2]

**TABLE I. Examples of Nucleophilic and Electrophilic Carbon–Carbon Bond-Forming Agents**

| Nucleophiles | Electrophiles |
|---|---|

## 5.1  CARBON–CARBON SINGLE-BOND FORMATION

### 5.1.1  Reactions in Basic Solution

Basic solutions generally have excess electron pairs available for coordination. This excess is usually prepared by introduction of a reducing agent. The more electropositive metals, in the metallic state, are strong reducing agents, that is, electron donors. They will react with most organohalides, reducing them to halide ion and negatively charged, strongly basic carbon. These carbons have a complete octet of electrons at the expense of the metals. The new carbon–metal bond is substantially ionic, and the reagents are referred to as *carbanions*, with the metals being essentially cations.[3] Magnesium, lithium,[4] zinc, and sometimes sodium and potassium are used in this way. These carbanions are powerful nucleophiles toward most electrophiles. In some cases the reactivity is lowered by exchanging less electropositive metal cations (such as copper, mercury, or cadmium) for those initially used in order to obtain selectivity on polyfunctional electrophiles. This direct use of reducing metals on organohalides is commonly the source of simple alkyl, aryl, and vinyl carbanions, and ester enolates.

Et₃N    K₂CO₃    NaOH    NaOC₂H₅    KOC(CH₃)₃ ˙ LiN(CH(CH₃)₂)₂    *n*-BuLi    *t*-BuLi
**Weakest**                                                                     **Strongest**

**Figure I.** Some bases suitable for preparing carbanions from their conjugate acids. The basicity of alkyllithium reagents is even higher in the presence of tetramethylethylenediamine (TMEDA).

The second routine method of preparing carbanions begins, not with organohalides, but with reagents where carbon carries a hydrogen of sufficient acidity to be removed by a stronger base. A selection of bases for this purpose is given in Fig. I.

The weakest bases are used to generate small equilibrium concentrations of carbanions that are often sufficient for high yield overall carbon–carbon bond formation. Sodium ethoxide will give nearly complete formation of carbanions that are resonance delocalized to two oxygens as in diethyl malonate anion. Ketone and ester enolates are usually prepared using lithium diisopropylamide (LDA) (Eq. 1).[5] This stronger base has

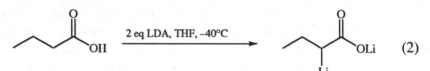

$$\underset{\text{CH}_3\text{COCH}_3}{\overset{\text{O}}{\parallel}} \xrightarrow{\text{LDA, THF, }-78°\text{C}} \left[ \underset{^-\text{CH}_2\text{COCH}_3}{\overset{\text{Li}^+ \; \text{O}}{\parallel}} \quad \underset{\text{CH}_2\text{=COCH}_3}{\overset{\text{Li}^+ \; \text{O}^-}{\;}} \quad \underset{\text{CH}_2\text{=COCH}_3}{\overset{\text{Li}\text{–O}}{\;}} \right] \quad (1)$$

little nucleophilicity because of the high steric hindrance of the isopropyl groups. LDA is strong enough to remove two protons from carboxylic acids, giving a dianion that has high reactivity toward electrophiles at the α position (Eq. 2).[6]

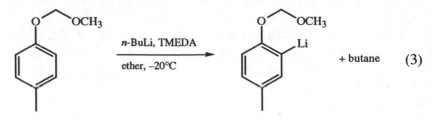

$$\text{2 eq LDA, THF, }-40°\text{C} \quad (2)$$

Alkyllithium reagents are basic enough to remove protons from aromatic ring carbons ortho to cation coordinating substituents as shown in Eq. 3.[7] Some other suitable coordinating substituents are shown in Fig. II.[8] In the amides and urethanes, steric hindrance is sufficient to avoid attack of the alkyllithium at the carbonyl carbon. The strongest bases,

*n*-BuLi, TMEDA

ether, −20°C

+ butane    (3)

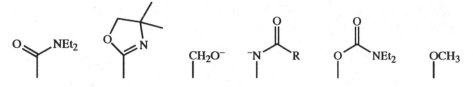

**Figure II.** Substituents that direct metallation to the ortho position on an aromatic ring.

the alkyllithiums, are also strongly nucleophilic and are usually limited to use with carbanion precursors that have no electrophilic carbonyl groups.

The stronger bases in Fig. I are again made by a redox reaction of a metal or by electrochemical reduction. Thus in this second method, carbanions are again prepared ultimately with a reducing agent, but a proton transfer step is included.

The relative acidities of these carbanion precursors depend on the extent of delocalization of the resulting negative charge in the structure. The factors include resonance to electronegative atoms, inductive stabilization by adjacent neutral or positively charged sulfur or phosphorus, and lower hybridization on the carbon itself.

A third method for preparation of carbanions is metal halogen exchange. An aryl or vinyl halide may be treated with an alkyllithium that gives the aryl or vinyllithium and the alkyl halide (Eq. 4).[9] Vinyl bro-

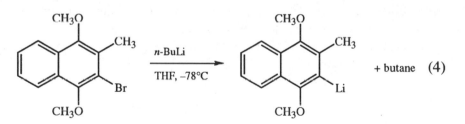

$$+ \text{butane} \quad (4)$$

mide can be converted to vinyllithium by treatment with *tert*-butyllithium in ether at $-78°C$.[10]

The carbanions from any of these methods are usually combined with electrophiles immediately after they are prepared or even during their preparation. The four common kinds of carbon–carbon bond-forming electrophiles give alkylation, acylation, addition, and conjugate addition.

Simple primary or secondary alkyl chlorides, bromides, iodides, or tosylates will give alkylation. For the reaction in Eq. 5, a weak base was used to generate a small concentration of the resonance delocalized enolate anion.[11] In this case the electrophile, methyl iodide, is present from the start. It approached the carbanion from the least hindered side to give the diastereomer shown.

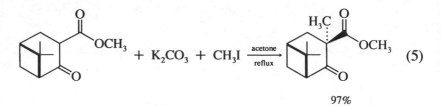

$$+ \ K_2CO_3 \ + \ CH_3I \ \xrightarrow[\text{reflux}]{\text{acetone}}$$

(5)

97%

The cyanide ion is only weakly basic and is thus available as salts that can be used in water solution. It can be alkylated most readily under phase-transfer conditions by using a quaternary ammonium catalyst (Eq. 6).[12] The alkylations also proceed well under homogeneous conditions in DMSO.

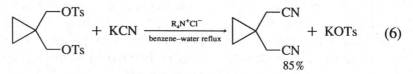

$$+ \ KCN \ \xrightarrow[\text{benzene–water reflux}]{R_4N^+Cl^-}$$

$$+ \ KOTs$$

(6)

85%

Trimethylsilyl ethers may serve as electrophiles and can be prepared *in situ* from alcohols. Heating an alcohol at 65°C with sodium cyanide, trimethylchlorosilane, and a catalytic amount of sodium iodide in acetonitrile–DMF gives the nitriles in a single operation.[13] Good yields are obtained with primary, secondary, and tertiary alcohols, and inversion of configuration has been demonstrated in a secondary case.

The alkylation of simple Grignard and organolithium compounds requires copper catalysis:[14]

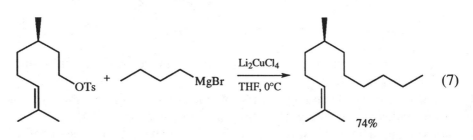

$$\xrightarrow[\text{THF, 0°C}]{Li_2CuCl_4}$$

(7)

74%

Carbanions will add to ketones and aldehydes to give alcohols. The carbanions prepared in Eqs. 1 and 2 were added to aldehydes as shown in Eqs. 8 and 9.[15]

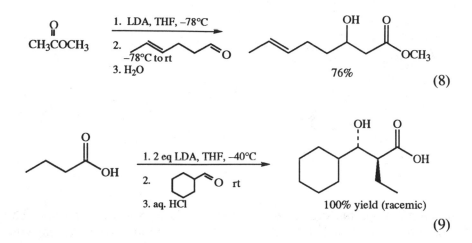

$$(8)$$

$$(9)$$

Organocuprates add at the $\beta$ position of $\alpha,\beta$-unsaturated aldehydes, ketones, esters, and *N,N*-dialkylamides. The resulting enolate anions may be captured in situ with trimethylchlorosilane and then hydrolyzed (Eq. 10).[16] The organocuprates are slow to react with trimethylchloro-

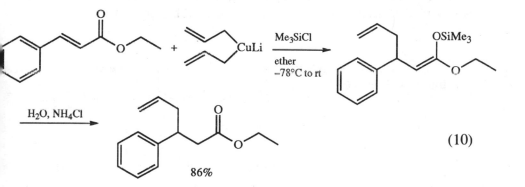

$$(10)$$

silane; therefore, they can be combined at $-78°C$ and then the $\alpha,\beta$-unsaturated electrophile added last. In this way the reactions are faster, 20% excess cuprate is sufficient, and the intermediate enolate is converted to the enol silyl ether before it can give by-products from competing aldol or Claisen reactions.[17]

Ketone enolates and highly resonance delocalized carbanions such as those from 1,3-dicarbonyl compounds will likewise bond to the $\beta$ position of $\alpha,\beta$-unsaturated ketones or esters to give an enolate anion that is protonated to give overall addition to the $\alpha,\beta$ unsaturation:

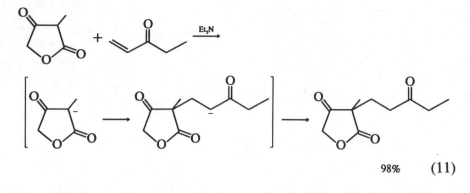

$$98\% \quad (11)$$

Such additions are called *conjugate addition* or *Michael addition*. The more reactive, less stabilized nucleophiles including Grignard reagents, alkyllithium reagents, and simple enolates from esters, nitriles, amides, or carboxylic $O,\alpha$-dianions usually do not give conjugate addition, but bond at the carbonyl carbon.

$\alpha,\beta$-Acetylenic esters and acetals undergo conjugate addition, also. If an organocuprate is added at low temperature and the resulting anion protonated also at low temperature, the process is a stereoselective syn addition of R and H as shown in Eq. 12.[18] This amounts to a stereo-

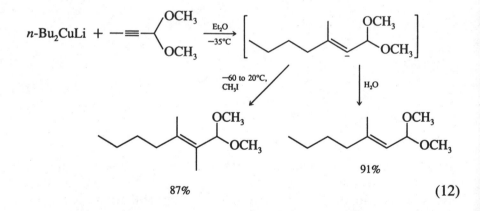

$$87\% \qquad\qquad 91\% \qquad\qquad (12)$$

selective synthesis of trisubstituted alkenes. If instead of protonating, the anion is alkylated, even tetrasubstituted alkenes are available stereospecifically. The requisite acetylenic acetals are readily made from the anions of the 1-alkynes plus methyl orthoformate. Esters may be used similarly, but the intermediate anion is less reactive.[19]

Acid chlorides, anhydrides, and esters react with carbanions to acylate them, affording ketones. An intramolecular example is shown in Eq.

13.[20] The base gave the lactone enolate, which was then acylated by the

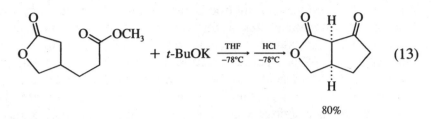

$$\qquad (13)$$

80%

methyl ester. Alkyl- or aryllithium and magnesium reagents tend to react twice with these acylating agents, leading to tertiary alcohols. This may be avoided by treating the Grignard reagent with an acid chloride in THF (not ether)[21] at −78°C, or by converting the organolithium reagent to the cuprate and then treating with the acid chloride at −78°C. The less reactive nitriles or tertiary amides will also acylate the Grignard reagents (Eq. 14).[22] The carbanion from Eq. 3 was formylated with DMF (Eq. 15). In each of these carbon–carbon bond-forming reactions,

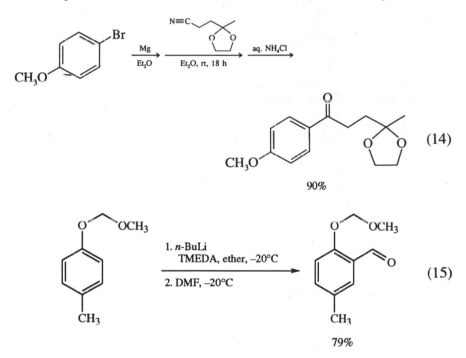

$$\qquad (14)$$

90%

$$\qquad (15)$$

79%

part of the driving force is the formation of a less basic anion in place of the carbanion, that is, a $Cl^-$ or $Br^-$ from alkylation and acylation, an alkoxide from addition or acylation, or an enolate from conjugate addition.

Ketones can present a problem in specificity. Under basic conditions they may react with two or more molecules of the electrophile to give a mixture of products. Furthermore, unsymmetric ketones may present a choice of two enolate sites so that control is necessary to direct to the desired one. Many alternatives have been developed for this problem. One solution is to incorporate a temporary group on one enolate site to render that site more acidic so that the electrophile will react there. The familiar β-ketoester reactions (acetoacetic ester synthesis) are widely used. For another alternative, the ketone is first converted to an imine (Section 4.2.3) or a dimethyl hydrazone, and the enolate of that derivative is used with electrophiles.[23] The enolate forms selectively on the least substituted α carbon and also gives selective monoalkylation or addition. In Eq. 16 we see an example of a directed aldol condensation.[24]

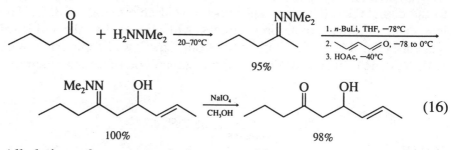

Alkylation of unsymmetric ketones without derivatization will give mostly reaction at the more substituted enolate site under reversible deprotonating conditions.

Aldehyde enolates present another problem. They tend to give self-condensation before an electrophile can be added. This may be solved again by use of imine enolates or N,N-dimethylhydrazones, which are themselves of low electrophilicity and allow good crossed aldol condensations and alkylations. For example, the tert-butyl imine of propanal was converted to the enolate with LDA and used in a crossed aldol condensation:[25]

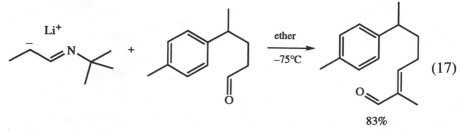

## 5.1.2  Reactions in Acidic Solution

Strong acids produce carbocations from a variety of functional molecules. Protonation of alcohols, epoxides, carbonyl compounds, and alkenes does so. Lewis acids such as anhydrous aluminum chloride can combine with the foregoing substrates and can also remove halide ions from carbon to give carbocations. Diazotization of primary amines in acid solution in another source.

The carbocations are transient intermediates, generated in the presence of alkene or arene nucleophiles to give carbon–carbon bond formation. This gives a new carbocation requiring a second step, which may be deprotonation, bonding to an oxygen or a halide ion, loss of a silyl group, or abstraction of a hydride. Carbocations may react with alkenes and dienes to give new carbocations, which may do likewise repetitively to give polymers.[26] Some varieties of synthetic rubber are produced in this way. Although $\sigma$-bonding electrons are more tightly held than $\pi$ electrons, they will react with carbocations particularly when they are arranged close by for intramolecular (rearrangement) processes. Migration of an atom or group with the $\sigma$-bonding pair from an adjacent carbon to the initial carbocationic site will occur if the new carbocation has greater stability (delocalization) from electron-donating alkyl groups, adjacent nonbonding electron pairs, or resonance to allylic sites, or if strain energy is released.

The example in Eq. 18[27] shows the formation of a carbocation from an epoxide, reaction with an alkene, and finally aromatic substitution on a furan.

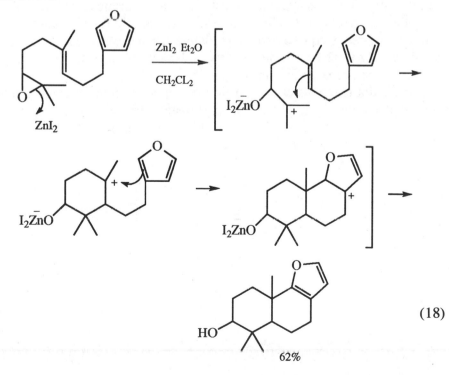

(18)

62%

Trimethylsilyl enol ethers are especially valuable nucleophiles toward carbocations.[28] After attachment of the carbocation, the trimethylsilyl group is readily removed by a halide ion to afford a ketone (Eqs. 19 and 20).[29,30] This is analogous to the alkylation of ketone enolate anions

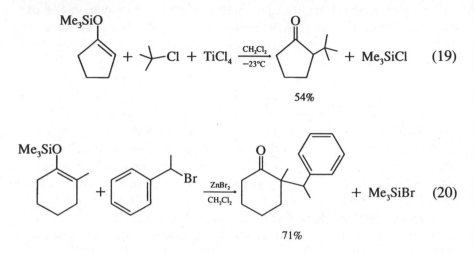

$$54\%$$

$$(19)$$

$$71\%$$

$$(20)$$

but differs in several ways. Here a specific enol ether (Section 4.2.5) can be used, restricting the alkylation to one site and giving no dialkylation (which sometimes competes in enolate anion alkylation). Most significantly, it allows attachment of tertiary alkyl groups and others that would have given mostly elimination in basic solutions. The reaction in Eq. 19 required 1 eq of Lewis acid, while the benzylic case in Eq. 20 required only 0.02 eq of catalyst. Even 2-methyl-2-*tert*-butylcyclohexanone can be prepared in this way in 48% yield.

Some aldol reactions can be carried out in acid. Here the nucleophile is an enol and the electrophile is the protonated carbonyl group. Equation 21 shows the cyclization of a keto aldehyde.[31] The acidic conditions generally give dehydration of the aldol.

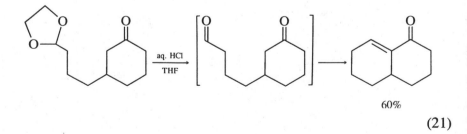

$$60\%$$

$$(21)$$

Crossed aldol condensations[32] between dissimilar ketones may be carried out under Lewis acid conditions using the silyl enol ether of that ketone intended as the nucleophile. This affords the aldols without de-hydration or polycondensation (Eq. 22).[33]

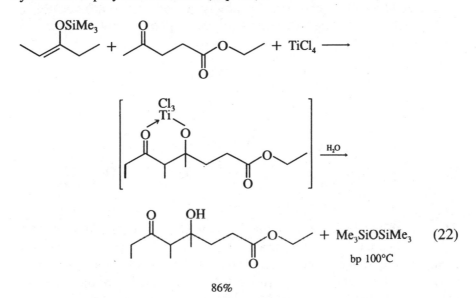

Simple ketones and 1,3-diketones give conjugate addition in acidic solution as shown in Eq. 23.[34] Here, too, the silyl enol ethers and $TiCl_4$ may be used (Eq. 24).[35]

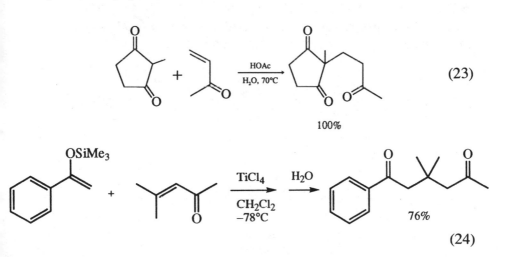

## 5.2 CARBON–CARBON DOUBLE-BOND FORMATION

Two carbons may be brought together and joined by a double bond. Typically the electrophilic side will be a ketone or an aldehyde. If the nucleophilic side is simply an alkylmagnesium halide or an alkyllithium, the secondary or tertiary alcohol intermediate may be dehydrated, but in many cases there is a choice of sites for the double bond and it may arise elsewhere than between the newly joined carbons. Even the β-hydroxyesters from the Reformatsky reaction often give mixtures of alkenes:[36]

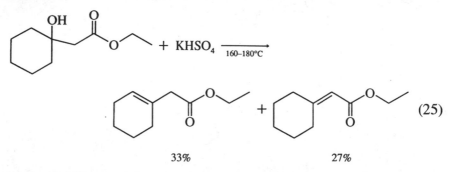

33%          27%

This variability may be prevented and the double bond formed specifically between the joining carbons if the nucleophilic carbon bears a group with a high oxygen affinity that will leave with the oxygen atom. That departing group takes the role of the departing $H^+$ of the preceding cases but is available at only one site; thus the specificity. Silicon and phosphorus are excellent in that role.

Trimethylsilylacetate esters may be converted to the enolate by treatment with lithium dialkylamide bases in THF at −78°C. These will add to ketones or aldehydes quickly at −78°C, followed by elimination of $Me_3SiOLi$ and formation of α,β-unsaturated esters in high yields, uncontaminated by β,γ-unsaturated isomers:[37,38]

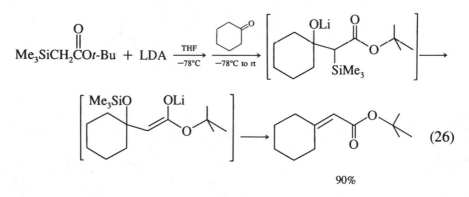

90%

The reaction mixture is quenched with aqueous HCl, extracted, and distilled. The by-product hexamethyldisiloxane, bp 100°C, is easily removed. This is known as the *Peterson reaction*.[39]

The requisite ethyl trimethylsilylacetate was made by the reaction of chlorotrimethylsilane, ethyl bromoacetate, and zinc.[40] It and the *tert-*

butyl ester can also be made by treating ethyl or *tert*-butyl acetate with LDA followed by chlorotrimethylsilane. Esters of longer-chain acids give mostly *O*-silylation under these conditions, but diphenylmethylchlorosilane gives *C*-silylation selectively. These diphenylmethylsilylated esters give the Peterson reaction as well:[41]

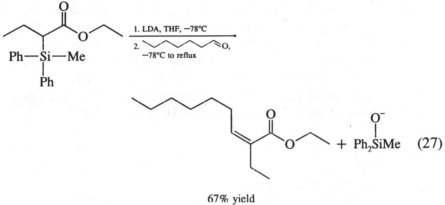

$$+ Ph_2SiMe \quad (27)$$

67% yield
78% Z, 22% E

Nonconjugated alkenes may be assembled by using a siloxide elimination, but the nucleophile is usually made in a different way since bases are unable to remove a proton alpha to a silicon without conjugative stabilization (unless it is a SiCH$_3$ site). Organolithium reagents will add to triphenylvinylsilane and may then be used with an aldehyde or ketone as exemplified by the synthesis of the alkene precursor of the sex pheromone of the gypsy moth:[42]

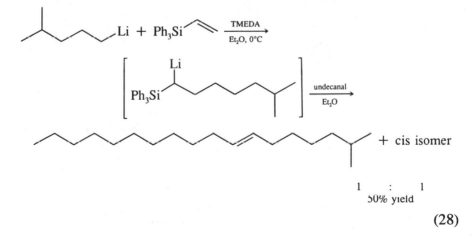

+ cis isomer

1  :  1
50% yield

(28)

Phosphorus has been used to a far greater extent for specific olefin synthesis.[43-45] Alkyl chlorides and bromides may be treated with triphenylphosphine to give quaternized salts. A base will remove a proton from a carbon alpha to the phosphorus to generate an ylide. The plus-charged phosphorus allows that proton removal in contrast to neutral silicon. Although the ylide carries no net charge, the substantial dipole gives high nucleophilic reactivity toward aldehydes and ketones to give an intermediate 1,2-oxaphosphetane that cleaves to the alkene and triphenylphosphine oxide. This is known as the *Wittig reaction*. The triphenyl phosphine oxide is nonvolatile and somewhat organic soluble and can be a nuisance to get rid of in comparison to hexamethyl disiloxane. In the absence of $Li^+$ ions, the reaction can give (Z)-stereoselectivity (Eq. 29)[46] up to 95%. The opposite stereoselectivity is obtained with Schlosser conditions,[47] where the diastereomeric intermediates are equilibrated with base before cleavage to alkene (Eq. 30).[46]

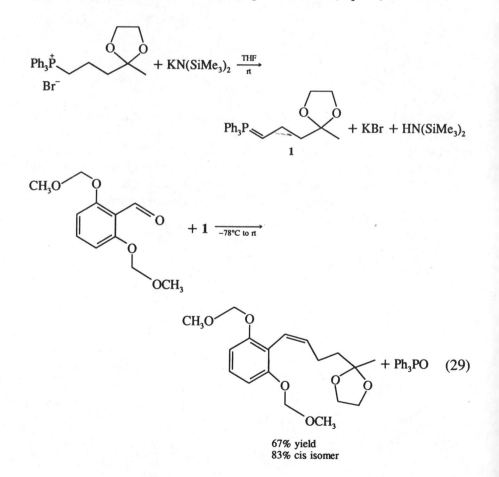

67% yield
83% cis isomer

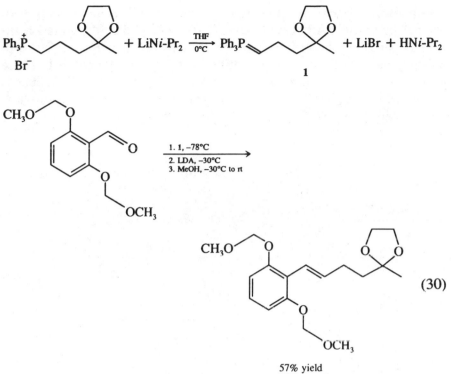

(30)

57% yield
95% trans isomer

$\alpha,\beta$-Unsaturated esters can be prepared using phosphonoesters.[48] In this case the leaving group with oxygen affinity is a phosphate, and the nucleophile is an enolate anion with stabilization by resonance with the ester carbonyl group and the phosphorous. The diethyl phosphate by-product is water-soluble and easily removed (Eq. 31).[49]

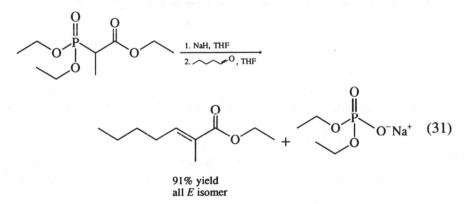

(31)

91% yield
all *E* isomer

The phosphonoesters are prepared by treating the $\alpha$-bromoesters with triethyl phosphite (Arbuzov reaction). As with ylides, the stereochemistry depends on the cations present. Aldehydes give $E$ products with ordinary bases such as sodium hydride as in Eq. 31. If a potassium base is used with crown ether complexation of the cation, the $Z$ product is favored. Even higher $Z$ selectivity is obtained using trifluoroethyl phosphonoesters (Eqs. 32, 33).[50]

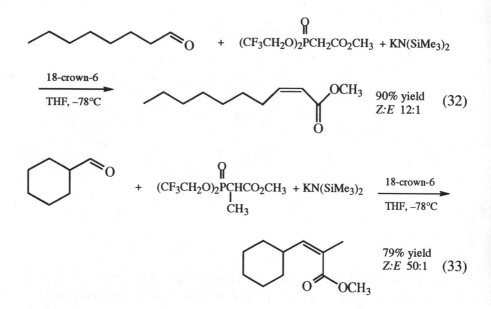

$$\text{90\% yield} \quad Z{:}E \ 12{:}1 \tag{32}$$

$$\text{79\% yield} \quad Z{:}E \ 50{:}1 \tag{33}$$

## 5.3  MULTIBOND PROCESSES

Many reactions result in a nearly simultaneous formation of a pair of $\sigma$ bonds. In some cases a carbene is a transient intermediate. Carbene $:CH_2$ is electron-deficient; it lacks two electrons for a complete octet. Although there is no net charge and little or no dipole, it is highly electrophilic and will attack both $\pi$ and $\sigma$ electrons to form pairs of new bonds. The lack of specificity in this high reactivity renders $:CH_2$ of little synthetic value, but dihalo carbenes are stabilized and selective toward alkenes. They give dihalocyclopropanes in good yield as shown in Section 9.6. Cyclopropanation of alkenes can also be accomplished via other electrophilic transient intermediates that are possibly metal complexes of carbenes. Copper, rhodium, or palladium catalyze the

decomposition of diazoketones or esters, which, in the presence of alkenes, gives cyclopropyl ketones or esters[51,52] as in Eq. 34.[53] Simple

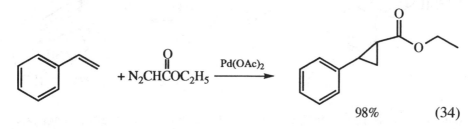

98%          (34)

cyclopropanation of alkenes may be accomplished by using dibromo-
methane, zinc dust, and copper(I) chloride promoted by a small amount
of acetyl chloride (Eq. 35).[54]

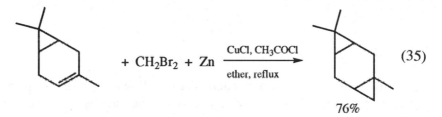

(35)

76%

Concerted reactions are commonly used to join carbons. For example,
the Diels–Alder reaction is the formation of a cyclohexene from a diene
and an alkene. Usually the alkene is rendered electrophilic by conjugation with a carbonyl group, and the diene may be rendered nucleophilic by electron-donating substituents. In the case shown in Eq. 36 the

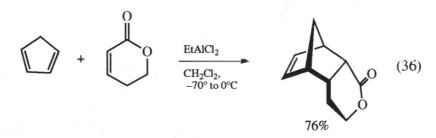

76%

alkene is further electron depleted by association with a Lewis acid,[55] a
common technique for accelerating Diels–Alder reactions. In some cases
the alkene is nucleophilic and the diene is electrophilic as in Eq. 37.[56]
Examples of this sort are called *reverse-electron-demand* Diels–Alder

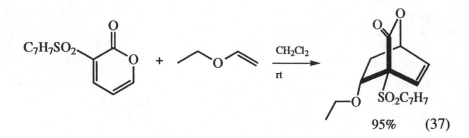

95%    (37)

reactions. It is important to point out here that the concerted reactions differ from the foregoing in that no carbanion or cation intermediate is involved, and in many cases electrophilic and nucleophilic factors are not present, as in the very favorable dimerization of cyclopentadiene. These reactions are covered in more detail in Chapter 8.

## PROBLEMS

Show how you would prepare each of the following products from the given starting materials. Where more than one step is required, show each step distinctly.

**1.**

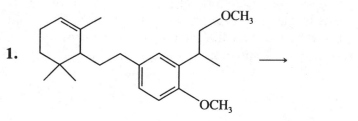

$\longrightarrow$

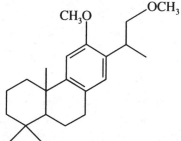

Ref. 57

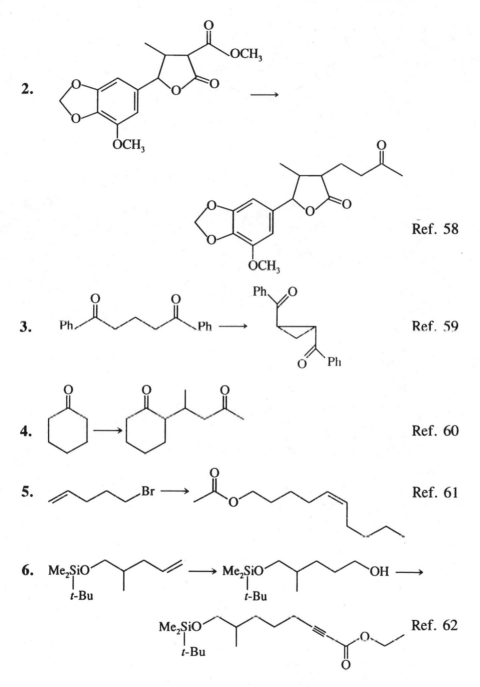

2.

Ref. 58

3.

Ref. 59

4.

Ref. 60

5.

Ref. 61

6.

Ref. 62

**7.**

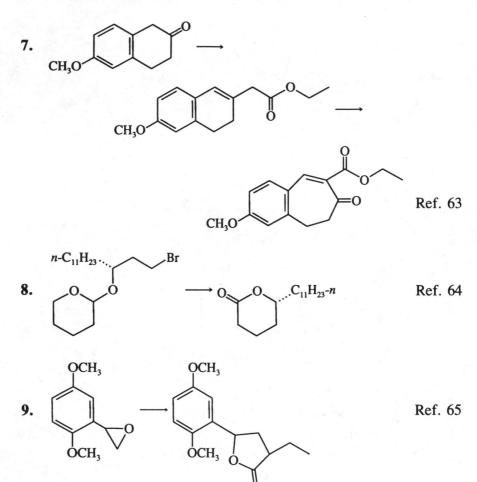

Ref. 63

**8.**

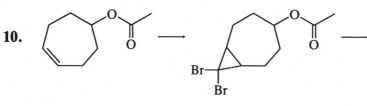

Ref. 64

**9.**

Ref. 65

**10.**

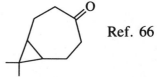

Ref. 66

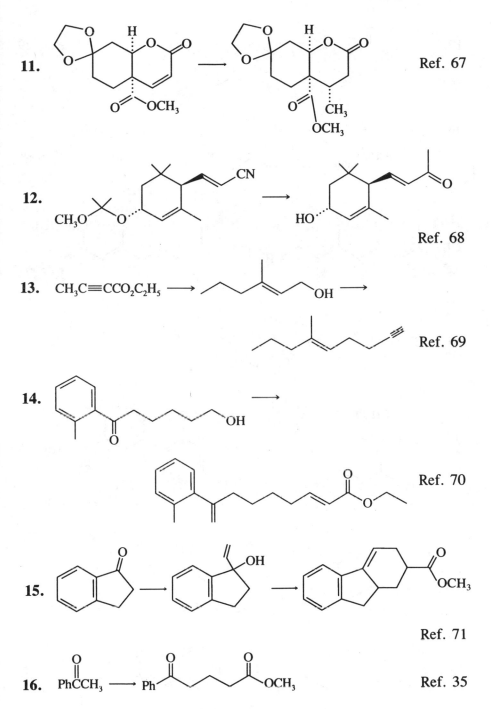

11.                                                                 Ref. 67

12.                                                                 Ref. 68

13.  CH₃C≡CCO₂C₂H₅ ⟶                    ⟶                          Ref. 69

14.                          ⟶                                     Ref. 70

15.                                                                Ref. 71

16.  PhCCH₃ ⟶ Ph                                                   Ref. 35

**17.** 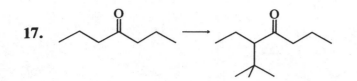 Ref. 29

**18.** 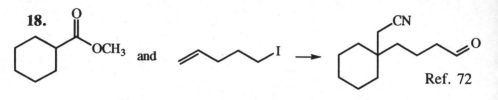 Ref. 72

**19.** 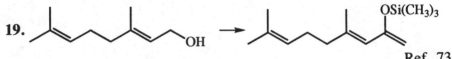 Ref. 73

**20.**  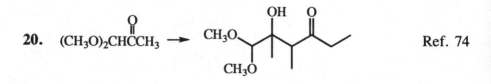 Ref. 74

$(CH_3O)_2CHCCH_3 \rightarrow$

**21.** 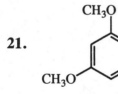 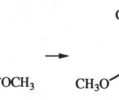 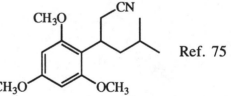 Ref. 75

**22.** 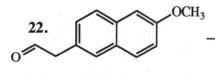 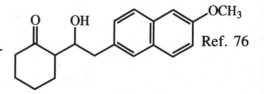 Ref. 76

**23.** 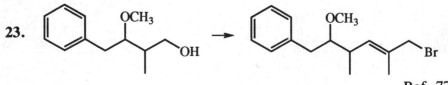 Ref. 77

**24.** 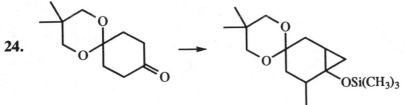    Ref. 78

**25.** 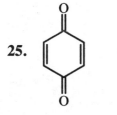    Ref. 79

**26.** 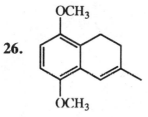    

**27.** 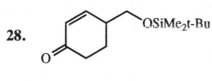    Ref. 80

**28.** 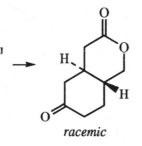    Ref. 81

racemic

**29.** 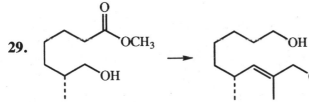    Ref. 82

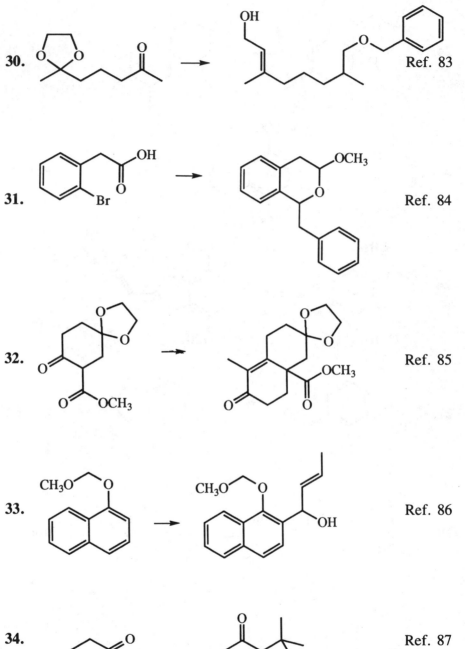

**30.** → Ref. 83

**31.** → Ref. 84

**32.** → Ref. 85

**33.** → Ref. 86

**34.** → Ref. 87

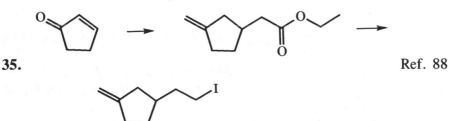

**35.**                                                                                 Ref. 88

## REFERENCES

1. Carruthers, W. *Some Modern Methods of Organic Synthesis*, 3rd ed., Cambridge Univ. Press, 1986.
2. Augustine, R. L., Ed. *Carbon–Carbon Bond Formation*, Vol. 1, Marcel Dekker, New York, 1979.
3. Stowell, J. C. *Carbanions in Organic Synthesis*, Wiley-Interscience, New York, 1979.
4. Wakefield, B. J. *Organolithium Methods*, Academic Press, New York, 1988.
5. Taber, D. F.; Amedio, J. C., Jr.; Raman, K. *J. Org. Chem.* **1988**, *53*, 2984.
6. Petragnani, N.; Yonashiro, M. *Synthesis.* **1982**, 521–578.
7. Harvey, R. G.; Cortez, C.; Ananthanarayan, T. P.; Schmolka, S. *J. Org. Chem.* **1988**, *53*, 3936.
8. Snieckus, V. *Chem Rev.* **1990**, *90*, 879.
9. Ghera, E.; Ben-David, Y. *J. Org. Chem.* **1988**, *53*, 2972.
10. Hecker, S. J.; Heathcock, C. H. *J. Org. Chem.* **1985**, *50*, 5159.
11. Inokuchi, T.; Asanuma, G.; Torii, S. *J. Org. Chem.* **1982**, *47*, 4622.
12. Foos, J.; Steel, F.; Rizvi, S. Q. A.; Fraenkel, G. *J. Org. Chem.* **1979**, *44*, 2522.
13. Davis, R.; Untch, K. G. *J. Org. Chem.* **1981**, *46*, 2985.
14. Raederstorff, D.; Shu, A. Y. L.; Thompson, J. E.; Djerassi, C. *J. Org. Chem.* **1987**, *52*, 2337.
15. Black, T. H.; DuBay, W. J., III; Tully, P. S. *J. Org. Chem.* **1988**, *53*, 5922.
16. Baek, D.-J.; Daniels, S. B.; Reed, P. E.; Katzenellenbogen, J. A. *J. Org. Chem.* **1989**, *54*, 3963.
17. Alexakis, A.; Berlan, J.; Besace, Y. *Tetrahedron Lett.* **1986**, *27*, 1047.
18. Alexakis, A.; Commercon, A.; Coulentianos, C.; Normant, J. F. *Tetrahedron* **1984**, *40*, 715.

19. Corey, E. J.; Katzenellenbogen, J. A. *J. Am. Chem. Soc.* **1969**, *91*, 1851.

20. Boeckman, R. K., Jr.; Naegley, P. C.; Arthur, S. D. *J. Org. Chem.* **1980**, *45*, 752.

21. Sato, F.; Inoue, M.; Oguro, K.; Sato, M. *Tetrahedron Lett.* **1979**, 4303.

22. Martin, S. F.; Puckette, T. A.; Colapret, J. A. *J. Org. Chem.* **1979**, *44*, 3391.

23. Whitesell, J. K.; Whitesell, M. A. *Synthesis*, **1983**, 517–536.

24. Corey, E. J.; Enders, D. *Chem. Ber.* **1978**, *111*, 1337, 1362.

25. Büchi, G.; Wüest, H. *J. Org. Chem.* **1969**, *34*, 1122.

26. Kennedy, J. P. *Cationic Polymerization of Olefins: A Critical Inventory*, Wiley-Interscience, New York, 1975.

27. Tanis, S. P.; Herrinton, P. M. *J. Org. Chem.* **1983**, *48*, 4572.

28. Brownbridge, P. *Synthesis* **1983**, 1.

29. Chan, T. H.; Paterson, I.; Pinsonnault, J. *Tetrahedron Lett.* **1977**, 4183.

30. Paterson, I. *Tetrahedron Lett.* **1979**, 1519.

31. Abbott, R. E.; Spencer, T. A. *J. Org. Chem.* **1980**, *45*, 5398.

32. Mukaiyama, T. *Org. React.* **1982**, *28*, 203–335.

33. Banno, K. *Bull. Chem. Soc. Jpn.* **1976**, *49*, 2284.

34. Hajos, Z. G.; Parrish, D. R. *Org. Synth.* **1985**, *63*, 26.

35. Narasaka, K.; Soai, K.; Aikawa, Y.; Mukaiyama, T. *Bull. Chem. Soc. Jpn.* **1976**, *49*, 779.

36. Kon, G. A.-R.; Nargund, K. S. *J. Chem. Soc.* **1932**, 2461.

37. Hartzell, S. L.; Sullivan, D. F.; Rathke, M. W. *Tetrahedron Lett.* **1974**, 1403.

38. Taguchi, H.; Katsuchi, S.; Yamamoto, H.; Nozaki, H. *Bull. Chem. Soc. Jpn.* **1974**, *47*, 2529.

39. Ager, D. J. *Org. Reactions*, **1990**, *38*, 1.

40. Fessenden, R. J.; Fessenden, J. S. *J. Org. Chem.* **1967**, *32*, 3535.

41. Larson, G. L.; Fernandez de Keifer, C.; Seda, R.; Torres, L. E.; Ramirez, J. R. *J. Org. Chem.* **1984**, *49*, 3385.

42. Chan, T. H.; Chang, E. *J. Org. Chem.* **1974**, *39*, 3264.

43. Maercker, A. *Org. React.* **1965**, *14*, 270.

44. Bestmann, H. J.; Vostrowski, O. *Topics Current Chem.* **1983**, *109*, 85.

45. Maryanoff, B. E.; Reitz, A. D. *Chem. Rev.* **1989**, *89*, 863–927.

46. Koreeda, M.; Hulin, B.; Yoshihara, M.; Townsend, C. A.; Christensen, S. B. *J. Org. Chem.* **1985**, *50*, 5426.

47. Schlosser, M.; Christman, K. F.; Piska, A. *Chem. Ber.* **1970**, *103*, 2814.

48. Wadsworth, W. S., Jr. *Org. React.* **1977**, *25*, 73.

49. White, J. D.; Takabe, K.; Prisbylla, M. P. *J. Org. Chem.* **1985**, *50*, 5233.

50. Still, W. C.; Gennari, C. *Tetrahedron Lett.* **1983,** *24*, 4405.

51. Dave, V.; Warnhoff, E. *Org. React.* **1970,** *18*, 217.

52. Barton, W. R.; DeCamp, M. R.; Hendrick, M. E.; Jones, M., Jr.; Levin, R.; Sohn, M. B. In *Carbenes*, Jones, M., Jr.; Moss, R. A. Eds., Wiley-Interscience, New York, 1973; Marchand, A. P.; MacBrockway, N. *Chem. Rev.* **1974,** *74*, 431.

53. Anciaux, A. J.; Hubert, A. J.; Noels, A. F.; Petiniot, N.; Teyssie, P. *J. Org. Chem.* **1980,** *45*, 695.

54. Fredrich, E. C.; Lewis, E. J. *J. Org. Chem.* **1990,** *55*, 2491.

55. Ikeda, T.; Yue, S.; Hutchinson, C. R. *J. Org. Chem.* **1985,** *50*, 5193.

56. Posner, G. H.; Wettlaufer, D. G. *Tetrahedron Lett.* **1986,** *27*, 667.

57. Matsumoto, T.; Imai, S.; Miuchi, S.; Sugibayashi, H. *Bull. Chem. Soc. Jpn.* **1985,** *58*, 340.

58. Matsumoto, T.; Imai, S.; Yamaguchi, T.; Morihira, M.; Murakami, M. *Bull. Chem. Soc. Jpn.* **1985,** *58*, 346.

59. Colon, I.; Griffin, G. W.; O'Connell, E. J., Jr. *Org. Synth.* **1972,** *52*, 33.

60. Huffman, J. W.; Potnis, S. M.; Satish, A. V. *J. Org. Chem.* **1985,** *50*, 4266.

61. Bestmann, H. J.; Vostrowski, O.; Koschatsky, K. H.; Platz, H.; Brosche, T.; Kantardjicw, I.; Rhinewald, M.; Knauf, W. *Angew. Chem. Internatl. Ed.* **1978,** *17*, 768.

62. Walba, D. M.; Stoudt, G. S. *J. Org. Chem.* **1983,** *48*, 5404.

63. McChesney, J. D.; Swanson, R. A. *J. Org. Chem.* **1982,** *47*, 5201.

64. Kikukawa, T.; Tai, A. *Chem. Lett.* **1984,** 1935.

65. Coburn, C. E.; Anderson, D. K.; Swenton, J. S. *J. Org. Chem.* **1983,** *48*, 1455.

66. Taylor, M. D.; Minaskanian, G.; Winzenberg, K. N.; Santone, P.; Smith, A. B., III *J. Org. Chem.* **1982,** *47*, 3960.

67. White, J. D.; Matsui, T.; Thomas, J. A. *J. Org. Chem.* **1981,** *46*, 3376.

68. Mayer, H.; Ruttimann, A. *Helv. Chim. Acta* **1980,** *63*, 1451.

69. Bowlus, S. B.; Katzenellenbogen, J. A. *Tetrahedron Lett.* **1973,** 1277.

70. Hornback, J. M.; Barrows, R. D. *J. Org. Chem.* **1983,** *48*, 90.

71. Ranu, B. C.; Sarkar, M.; Chakraborti, P. C.; Ghatak, U. R. *J. Chem. Soc. Perkin Trans. I* **1982,** 865.

72. Callahan, J. F.; Newlander, K. A.; Bryan, H. G.; Huffman, W. F.; Moore, M. L.; Yim, N. C. F. *J. Org. Chem.* **1988,** *53*, 1527.

73. Kraus, G. A.; Hon, Y.-S.; Sy, J.; Raggon, J. *J. Org. Chem.* **1988,** *53*, 1397.

74. Tanabe, Y.; Ohno, N. *J. Org. Chem.* **1988,** *53*, 1560.

75. Lal, K.; Zarate, E. A.; Youngs, W. J.; Salomon, R. G. *J. Org. Chem.* **1988,** *53*, 3673.

76. Pataki, J.; Di Raddo, P.; Harvey, R. G. *J. Org. Chem.* **1989,** *54*, 840.

77. Namikoshi, M.; Rinehart, K. L.; Dahlem, A. M.; Beasley, V. R.; Carmichael, W. W. *Tetrahedron Lett.* **1989,** *30*, 4349.

78. Piers, E.; Friesen, R. W. *J. Org. Chem.* **1986,** *51*, 3405.

79. Flynn, G. A.; Vaal, M. J.; Stewart, K. T.; Wenstrup, D. L.; Beight, D. W.; Bohme, E. H. *J. Org. Chem.* **1984,** *49*, 2252.

80. Jardon, P. W.; Vickery, E. H.; Pahler, L. F.; Pourahmady, N.; Mains, G.; Eisenbraun, E. J. *J. Org. Chem.* **1984,** *49*, 2130.

81. Godleski, S. A.; Villhauer, E. B. *J. Org. Chem.* **1984,** *49*, 2246.

82. Ley, S. V.; Maw, G. N.; Trudell, M. L. *Tetrahedron Lett.* **1990,** *31*, 5521.

83. Hajos, Z. G.; Wachter, M. P.; Werblood, H. M.; Adams, R. E. *J. Org. Chem.* **1984,** *49*, 2600.

84. Harmata, M.; Barnes, C. L. *Tetrahedron Lett.* **1990,** *31*, 1825.

85. Pariza, R. J.; Fuchs, P. L. *J. Org. Chem.* **1983,** *48*, 2306.

86. Harvey, R. G.; Hahn, J.-T.; Bukowska, M.; Jackson, H. *J. Org. Chem.* **1990,** *55*, 6161.

87. Zambias, R. A.; Caldwell, C. G.; Kopka, I. E.; Hammond, M. L. *J. Org. Chem.* **1988,** *53*, 4135.

88. McMurry, J. E.; Andrus, W. A.; Musser, J. H. *Synthetic. Commun.* **1978,** *8*, 53.

# 6

# PLANNING MULTISTEP SYNTHESES

The challenge in synthesis is to devise a set of reactions that will convert inexpensive, readily available materials into complex, valuable products. Ordinarily this is not an obvious following of a roadmap, but rather a complex puzzle requiring much strategy. This chapter gives samples of the planning process with actual syntheses of relatively simple cases. Enough of the procedure is provided to enable you to analyze and devise syntheses for many molecules. A more extensive treatment is given in an excellent book by Warren.[1]

## 6.1  RETROSYNTHETIC ANALYSIS

You should familiarize yourself with what sorts of compounds are readily available by perusing commercial catalogs, but the actual process begins at the end of the synthesis; that is, you must study the desired structure and work *backward*. What penultimate intermediate would be readily convertible to that product, and then what before that? This process is called *retrosynthetic analysis*, and each backward step is indicated by a double-shafted arrow ($\Rightarrow$). With this a backward scheme is drawn, and then a forward process is developed with actual reagents, indicated with ordinary arrows. In more complicated syntheses you will need to look ahead toward steps in the middle of the process, but still a backward approach is most practical.

The steps include functional group interconversions as given in Chapter 4 and carbon backbone construction as illustrated in Chapter 5. Viewed as the disassembly of the product, you should first disconnect the parts that are joined by functional groups; for example, esters should be separated to acid and alcohol parts. The carbon–carbon bonds should be disconnected at or near functional groups and at branch points in the backbone. There are often a great many choices of dividing points and starting materials. For example, jasmone and dihydrojasmone have been made by hundreds of routes.[2] In selecting among choices, the number of steps should be minimized, cheaper starting materials selected, and high-yield reactions favored, and the scheme should converge instead of following a long linear sequence of steps. Sometimes a closely related molecule will be available, requiring a minimum of construction effort, as in making other steroids from diosgenin.

## 6.2 DISCONNECTION AT A FUNCTIONAL GROUP OR BRANCH POINT

Carbon–carbon bonds are frequently built by using carbonyl compounds. A carbonyl group normally confers a pattern of alternating potential electrophilic or nucleophilic reactivity along a carbon chain as shown in structure **1**.[3] The electrophilic character exists in the carbonyl com-

$$
\begin{array}{c}
\overset{\displaystyle O}{\underset{\displaystyle \|}{}} \\
\ldots C\!-\!C\!-\!C\!-\!C\!-\!C\!- \\
(+)\ (-)\ (+)\ (-)\ (+) \\
\mathbf{1}
\end{array}
$$

pounds themselves, continuing along the chain as far as conjugating $p$ orbitals are present to transmit it:

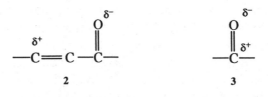

**2**          **3**

The nucleophilic character exists in the derived enolate **(4, 5)** or enol form:

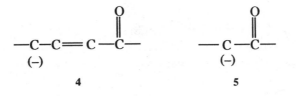

4                                    5

The charges at these sites serve to attract another carbon reagent of opposite charge and give a new bond.

If we consider a disconnection somewhere along the chain, we can decide whether the reactive site backed by the carbonyl group will be nucleophilic or electrophilic. Three different choices are taken in the following examples to illustrate the rationale.

Compound **6** is an intermediate in a synthesis of hemlock alkaloids. The carbamate functional group is made from a chlorocarbonate and the amine; therefore, that disconnection is the first retro step (Scheme I).

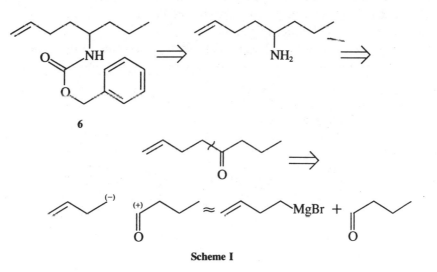

6

**Scheme I**

Since amines are often made from ketones, we go on to that key inter-mediate. Disconnection of the $\alpha$ carbon gives fragments with a (+) on the carbonyl (in accord with **1**) and requiring a (−) on the other reagent. We can now write a forward scheme with commercially available start-ing materials:[4]

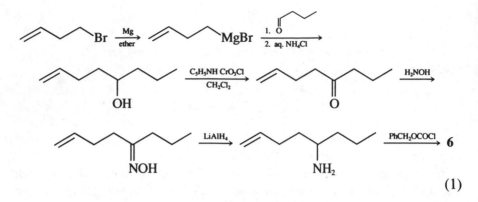

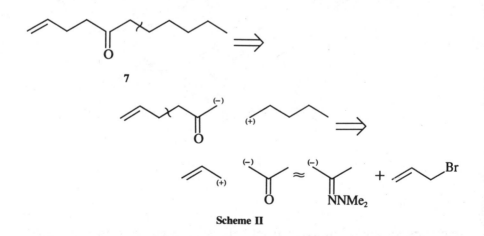

(1)

The 4-bromo-1-butene is available or can be prepared from the alcohol, which may, in turn, be prepared from vinylmagnesium chloride plus ethylene oxide. The Grignard reagent may be treated with an acylating agent, or as these authors chose, an aldehyde followed by an oxidizing agent. In this example the amine functional group suggested a carbonyl intermediate. This same retro step should be suggested by many other functional groups, including alcohols and halides.

A very similar ketone (**7**) was made with a disconnection between the $\alpha$ and $\beta$ carbons, in fact on both sides (Scheme II). In this disconnection

**Scheme II**

the carbonyl fragment is the nucleophile, specifically, the enolate of a hydrazone derivative (Section 5.1.1). The synthesis is shown in Eq. 2.[5]

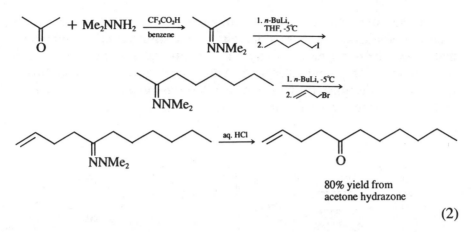

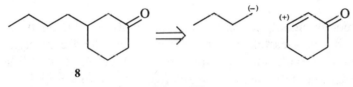

80% yield from
acetone hydrazone

(2)

Disconnection one atom further, that is between the $\beta$ and $\gamma$ carbons, requires a conjugate addition as in the synthesis of **8** (Scheme III). The

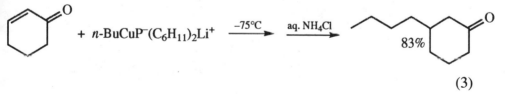

**8**

**Scheme III**

actual reaction is illustrated in Eq. 3.[6] Lithium di-*n*-butylcuprate is usually used in large excess, but the dicyclohexylphosphide reagent is not required in excess.

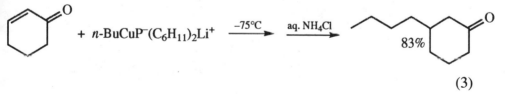

(3)

Which one of the three disconnections we select depends on other structural features of the particular product and on availability of materials. In the synthesis of **8** the other two disconnection choices would have required more steps and difunctional intermediate compounds since the compound is cyclic. Compounds **6** and **7** could be made by any of the three methods.

Under certain circumstances the normal electrophilic or nucleophilic sites in carbonyl compounds are unusable or do not give the easiest routes to products. For these situations we substitute another compound that allows the opposite reactivity but can subsequently be converted to the carbonyl compound. Reagents with this reversed or "abnormal" reactivity are designated by the German term *umpolung*.[7] The abnormal carbonyl α disconnection step is shown in Scheme IV. A carbonyl car-

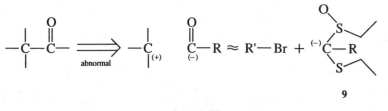

9

**Scheme IV**

bon is not nucleophilic, but several derivatives can be used that are nucleophilic at a carbon that can be converted to a carbonyl.[8] The dithioacetal monoxide anion **9** is an example. A preparation of 3-heptanone (Eq. 4) demonstrates the use of this reagent.[9] Some umpolung reagents

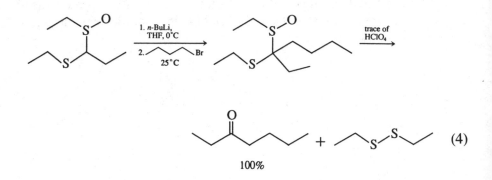

100%

(4)

are illustrated in Table I. Disconnecting one bond farther from the carbonyl, the α,β abnormal case is provided indirectly by the reaction of Grignard reagents with epoxides. This is followed by oxidation if the carbonyl group is the desired functionality. The β,γ disconnection with abnormal polarity is readily arranged if the α carbon is $sp^3$-hybridized and insulates the carbonyl group from the β carbon. In this case the

**TABLE I. Electrophilic and Nucleophilic Umpolung Reagents and Their Equivalents**

| | | |
|---|---|---|
| (structure) | $(-)\overset{O}{\overset{\|}{C}}-H$ | Ch. 6, Eq. 4 |
| $O_2N\diagup\diagdown R$ | $(-)\overset{O}{\overset{\|}{C}}-R$ | Ch. 6, Eq. 10 |
| (epoxide) R | $(+)\ C-\overset{O}{\overset{\|}{C}}-R$ | Ch. 6, Eq. 25 |
| Br (allyl bromide) | $(+)\ C-\overset{O}{\overset{\|}{C}}-R$ | Ch. 6, Eq. 11 |
| Br (ester) OR | $(+)\ C-\overset{O}{\overset{\|}{C}}-OR$ | Ch. 6, Eq. 12 |
| BrMg (dioxane) | $(-)\ C-C-\overset{O}{\overset{\|}{C}}-H$ | Ch. 6, Eqs. 5 and 16 |

carbonyl group must be protected from the nucleophilic $\beta$ carbon:[10]

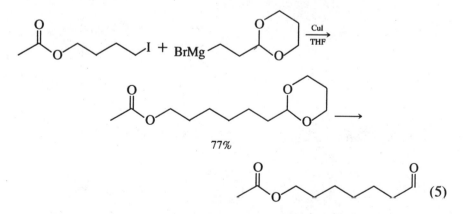

77%

(5)

Next we use these choices in some longer sequences. *exo*-Brevicomin
**(10)** is a pheromone from the Western pine beetle. Examining the func-

tionality, we see a carbon attached to two oxygens, that is, an acetal derivative of a ketone. Dividing this functionality to the components, we find a ketodiol. 1,2-Diols are usually made from alkenes. Bellas and co-workers[11] chose to make the double bond by the Wittig reaction and to use an $\alpha,\beta$ disconnection on the ketone (Scheme V).

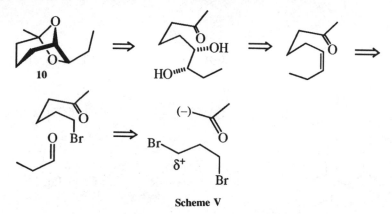

**Scheme V**

Of the various ways available for controlling monoalkylation of acetone, they chose to use the acetoacetic ester synthesis (Eq. 6). Notice also that the exo stereochemistry of the pheromone requires the threo diol. This could be made by syn glycolization of the trans alkene or by anti glycolization of the cis alkene. The latter was used here.

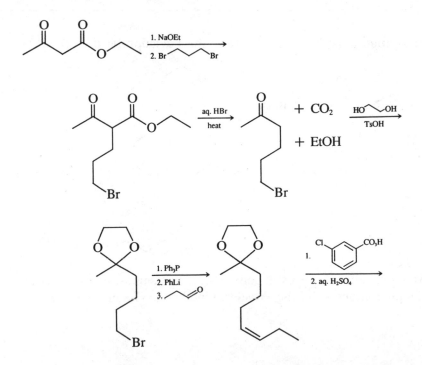

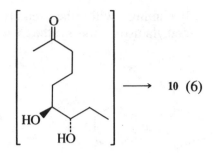

$$10 \quad (6)$$

7-Methoxy-α-calacorene **(11)** contains several branch points (carbons with three other carbons directly attached) where we may consider disconnecting with the help of temporary functional groups. To aid in choosing a retro starting point, we should look over the whole structure and consider steps that may be required. This structure includes a benzene ring that has other alkyl carbons attached forming another ring. The Friedel–Crafts acylation reaction would be a way to assemble this, using the carbonyl groups to incorporate the methyl and isopropyl groups. The methoxy group is a strong ortho, para director; therefore, the attachment at the para position should be developed first. This idea is elaborated in Scheme VI.

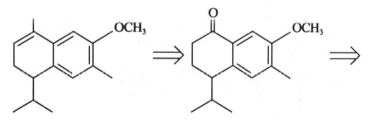

**11**

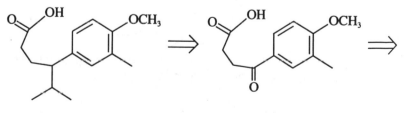

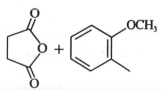

**Scheme VI**

Beginning with the readily available succinic anhydride and 2-methylanisole, the synthesis was carried out as in Eq. 7.[12] The ke-

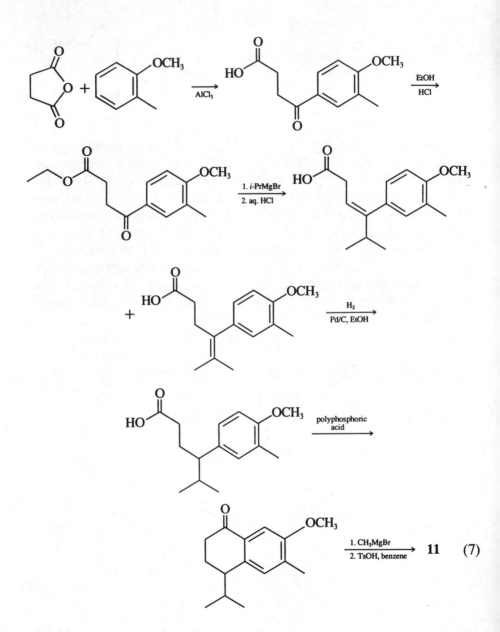

toester intermediate is sufficiently more reactive at the keto group to give the desired product from the Grignard treatment.

## 6.3  COOPERATION FOR DIFUNCTIONALITY

Molecules that contain two functional groups at particular distances apart
are assembled considering the electronic influence of both groups together. As with the monofunctional compounds, consider carbonyl
groups to be primary intermediates and examine their influence on the
fragments from disconnection between the groups. Beginning with a 1,3-
difunctional chain, we find fragments **12** and **13.** The normal influence

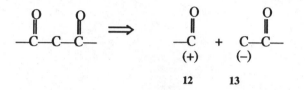

12        13

of the carbonyl in each fragment provides the charges shown. Since they
are opposite, they will attract each other, and this would be a favorable
approach to the synthesis of 1,3-difunctional compounds. To be specific,
consider the synthesis of **14.** This hydroxyester may be disconnected
between the α and β carbons to give appropriate fragments (Scheme
VII). The actual reagents could be the ester enolate in the form of a

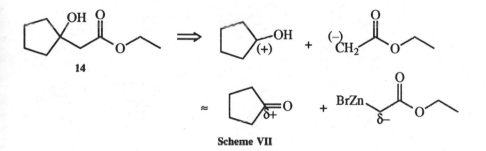

**Scheme VII**

Reformatsky reagent and the ketone. β Hydroxycarbonyl compounds are
often dehydrated; thus the α, β-, or β,γ-unsaturated compounds should
be approached in planning by first rehydrating. The initial adduct could
also be reduced or oxidized or further converted to other functional
groups; therefore, compounds **15** and **16** would also be approached with

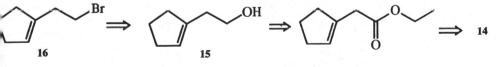

16            15

the same intermediates and disconnection in mind. The actual synthesis of **16** was carried out as in Eq. 8.[13] The dehydration gives the endocyclic double bond in five-membered or smaller rings.

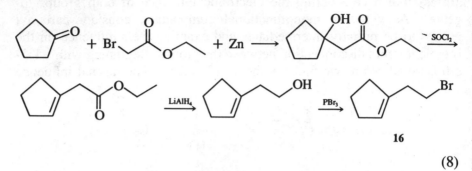

(8)

Bromoalcohol **17** is 1,3-difunctional, also. Maximum simplification would result from disconnecting the $C_3$ alcohol fragment, but first a carbonyl function should probably be considered. This gives a β-hydroxyester, so again the Reformatsky reaction is suggested (Scheme VIII). Equation 9 shows the sequence used[14] to prepare **17**.

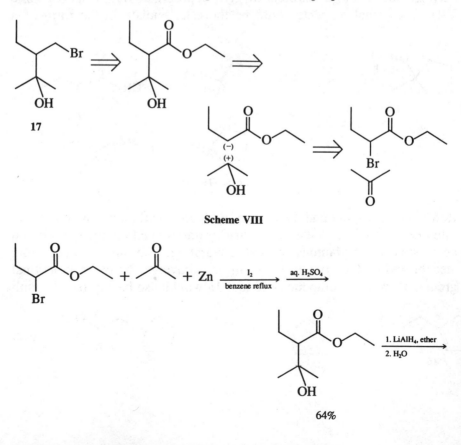

**Scheme VIII**

64%

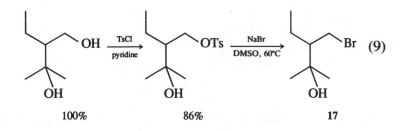

100%                    86%                    **17**

Turning next to a 1,4-difunctional chain, we find that neither of the two modes of disconnection give mutually attracting fragments (Scheme IX). It is therefore necessary to use an umpolung (reverse polarity)

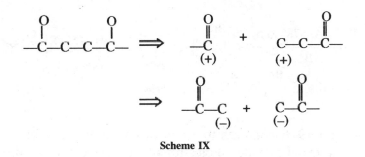

**Scheme IX**

reagent in place of one of the usual components. A nitro group on a carbon facilitates the removal of a proton from that carbon, giving a nucleophilic nitronate anion suitable for the 1,2 disconnection. Later the nitro group and carbon may be converted to a carbonyl group (which itself would have been electrophilic). 2,5-Heptanedione **(18)** was prepared in this way:[15]

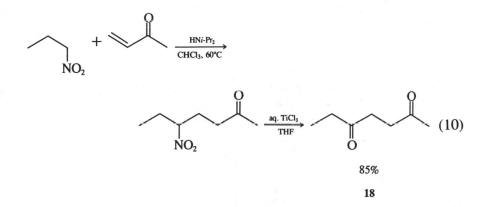

(10)

85%

**18**

The 2,3 disconnection of 1,4-difunctional molecules is also of value. The diketone in Scheme X disconnects to a pair of mutually unattractive

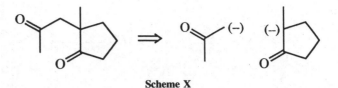

**Scheme X**

enolate ions, but one may be replaced with an umpolung reagent. 2,3-Dibromopropene is equivalent to acetone with an electrophilic α carbon. The actual steps are shown in Eq. 11. The initial ketone enolate was stirred at room temperature to equilibrate to the more stable more substituted enolate. The vinyl bromide intermediate was hydrolyzed to the ketone using mercury(II) acetate in formic acid:[16]

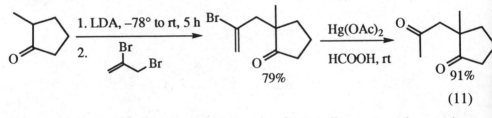

$$(11)$$

γ-Ketoester **19** disconnects to a pair of mutually unattractive enolate ions, but one may be rendered electrophilic by placing a bromine on the α carbon (Scheme XI). Thus the α-bromoester is an umpolung reagent. The retrosynthetic analysis of **19** may be continued to **20** by applying the principles already developed for 1,3-difunctional cases.

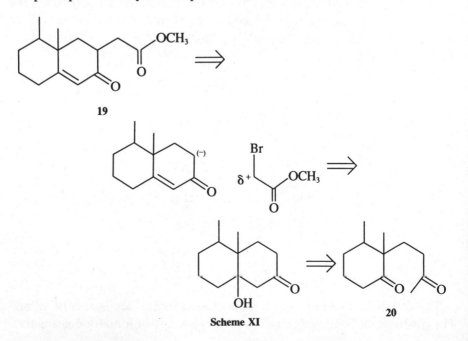

**Scheme XI**

Disconnections of 1,5-difunctional chains give normal charges that are favorable for mutual attraction of the fragments. A 2,3-disconnection (Scheme XII) gives components that could be an enolate ion and an $\alpha,\beta$-unsaturated carbonyl compound suitable for a conjugate addition.

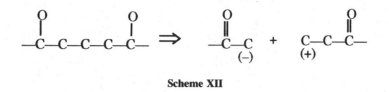

**Scheme XII**

The retrosynthetic analysis of **20** continues this way. We may expect 2,3-dimethylcyclohexanone to give enolate or enol preferentially at carbon 2. Equation 12 shows this step under acid conditions and also the continuation to compound **19**.[17]

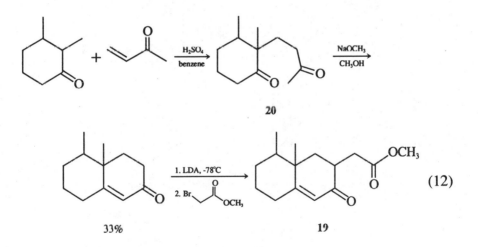

(12)

33%                                           **19**

Reviewing the reactive character of the fragments of disconnection of difunctional molecules, we can see a pattern. The 1,3 and 1,5 molecules give normally attractive fragments, while the 1,2 and 1,4 molecules require an umpolung reagent. Favorable and unfavorable normal charges alternate with increasing separation of the functional groups.

Lactone **21** is a 1,5-difunctional molecule at a lower oxidation state than **20,** and is accessible by the same disconnection rationale (Scheme XIII). First disconnect the functionality and then the carbon chain at the branch point (3,4 bond). The butyraldehyde enolate called for here is not practical because of self-condensation (Section 5.1.1); therefore, a nonelectrophilic derivative of the aldehyde is used instead. The piperi-

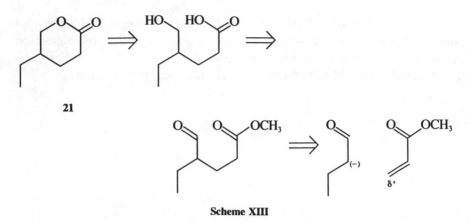

**21**

**Scheme XIII**

dine enamine is sufficiently nucleophilic for the Michael addition and can be hydrolyzed readily after that step (Eq. 13).[18]

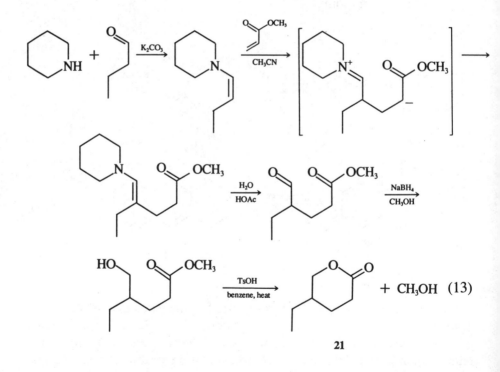

+ CH$_3$OH   (13)

**21**

1,6-Difunctional molecules are less often formed by connecting intervening carbons. The ready availability of cyclohexenes and cyclohexanones allows oxidative ring opening to give the 1,6-functionality spacing as exemplified in Eqs. 14[19] and 15.[20,21] Some 1,5-difunctional

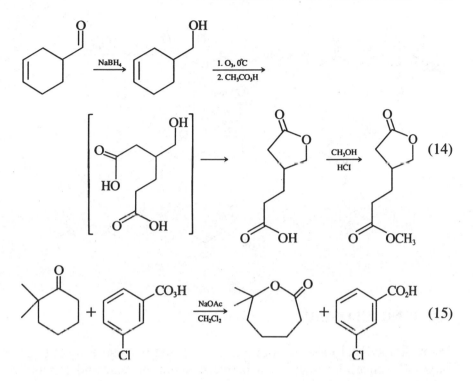

$$(14)$$

$$(15)$$

molecules are also formed by oxidative opening of five-membered ring compounds.

Difunctional molecules are sometimes assembled, ignoring the polarizing possibilities of one of the functional groups. Nucleophilic character can be developed at a site on a chain where it is insulated from a carbonyl group by intervening $sp^3$ carbons. The aldehydes or ketones then need protection as acetals. Equations 16 and 17 show preparation of 1,4 and 1,5 difunctionality this way.[22,23]

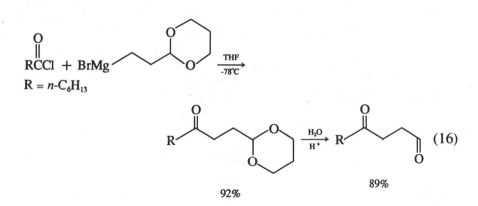

$$(16)$$

92%

89%

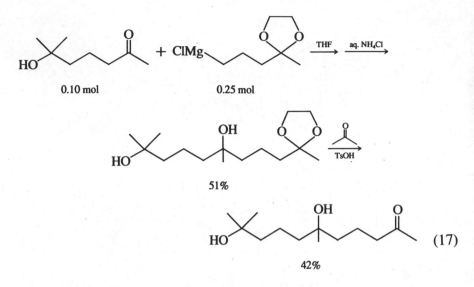

51%

(17)

42%

## 6.4  RING CLOSURE

Many difunctional molecules in this chapter and Chapter 5 react to give rings. A competition may exist between intermolecular and intramolecular reactions. In most cases the formation of three-, five-, and six-membered rings is more favorable than polymerization because the intermolecular process requires a bimolecular collision while cyclization requires only conformational alignment. This becomes less probable for rings larger than six; therefore, high dilution is used to lessen the frequency of intermolecular collisions. The acylion condensation[24] is efficient for such large-ring closures.

In some cases there is a competition within a molecule for closure to rings of different sizes. If the closure is an irreversible reaction such as alkylation of an enolate, kinetic control may allow three-membered ring closure to predominate over five-membered as in Eq. 18.[25] The electro-

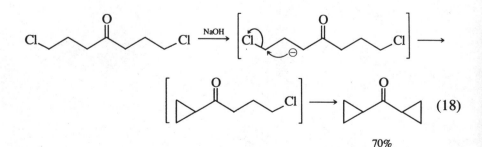

(18)

70%

philic carbon closer to the nucleophilic site is found first, in spite of the incorporation of ring strain. With reversible reactions, the thermodynamic product predominates. That is, the low strain five- or six-membered rings will form to the exclusion of three- or four-membered alternatives (Scheme XIV). In Claisen and aldol cyclizations, five- and

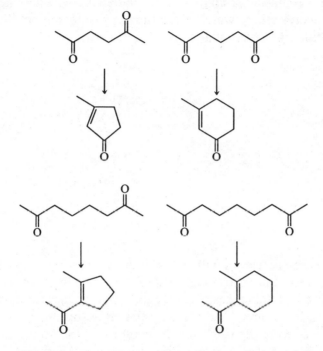

**Scheme XIV.**   Products expected from aldol cyclizations of 1,4- to 1,7-diketones.

six-membered rings are not in competition with each other. Seven-membered rings may sometimes be closed readily without high dilution:[26]

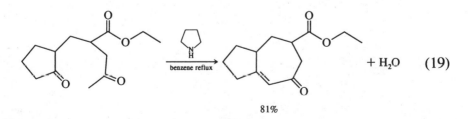

$$+ H_2O \qquad (19)$$

81%

Heterocyclic rings show the same size preferences. Halohydrins and base give epoxides under irreversible conditions. Hemiacetals, acetals, and lactones under reversible conditions favor five- or six-membered rings, where conformational flexibility permits, over intermolecular polymerization.

The closure of four-membered rings requires special methods. Three reactions are frequently used: the acyloin,[24] photochemical cycloaddition, and thermal ketene cycloadditions. Treatment of a 1,4-diester (Eq. 20)[27] with sodium metal and chlorotrimethylsilane in toluene gives the enediol bis-silyl ether, which is methanolyzed to the acylion in good yield. Without the chlorosilane, yields are poor.

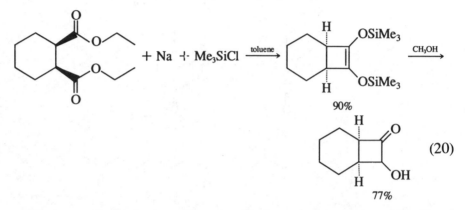

(20)

Alkenes and acetylenes will cycloadd photochemically to other alkene molecules, especially those conjugated to carbonyl groups, to give cyclobutanes or cyclobutenes.[28] The molecules are raised to an excited electronic state, sometimes via a radiation-absorbing sensitizer compound, add to form the ring, and descend to the electronic ground state (Section 8.3.2). In doubly unsymmetric cases the regio- and stereochemistry can be complex and dependent on conditions. Nevertheless, many are synthetically useful. A few examples are shown in Eqs. 21–23.[29–31]

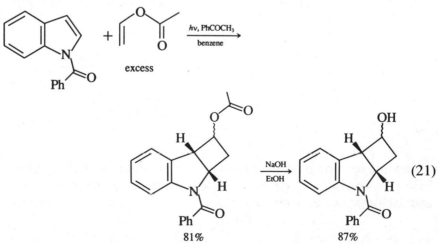

(21)

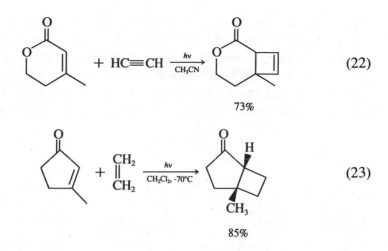

$$(22)$$

73%

$$(23)$$

85%

Ketenes $R_2C{=}C{=}O$ add thermally to alkenes to give cyclobuta-nones. Dichloroketene is readily generated *in situ* from trichloroacetyl chloride and copper-activated zinc metal. In Eq. 24 a silyl enol ether was treated with dichloroketene to give the four-membered ring regio-selectively. The chlorine atoms were then removed reductively with tri-*n*-butyltin hydride.[32]

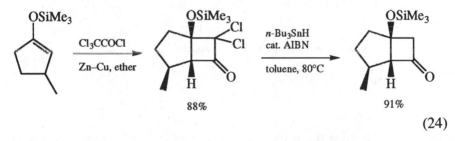

88%                                    91%

$$(24)$$

## 6.5 ACETYLIDE ALKYLATION AND ADDITION

In the remainder of this chapter, particular reactions are selected for examination of their synthetic potential. Acetylide ions are useful for linking carbon chains, particularly where a double bond is desired with stereoselectivity. Acetylene and 1-alkynes may be deprotonated with strong bases such as LDA and then treated with alkyl halides or carbonyl compounds. Preformed lithium acetylide complexed with ethylene-diamine is available as a dry powder. Several alkynes derived from acetylide and carbon dioxide or formaldehyde are available, including propargyl alcohol ($HC{\equiv}CCH_2OH$), propargyl bromide ($HC{\equiv}CCH_2Br$) and methyl propiolate ($HC{\equiv}CCO_2CH_3$).

A disconnection between a double bond and an allylic carbon should suggest acetylide chemistry, while disconnection between the double-bonded carbons should suggest the Wittig and allied reactions (Section 5.2). Consider structure **22.** The cis double bond could come from an acetylene, in fact, propargyl alcohol. According to retrosynthetic analysis, the amide should be disconnected first. The secondary amine could come from benzylamine and a halide, and that halide could be made via propargyl alcohol and ethylene oxide (Scheme XV). This entails acces-

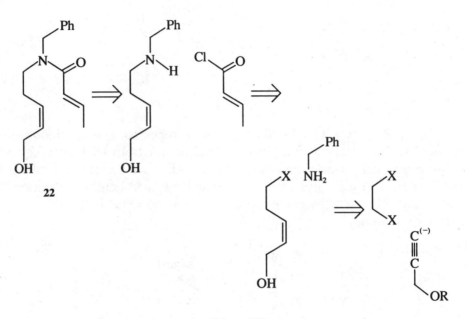

**Scheme XV**

sory steps to reduce the triple bond to a double bond and to distinguish the propargylic alcohol from the alcohol arising from the ethylene oxide opening. The actual synthesis is shown in Eq. 25.[33]

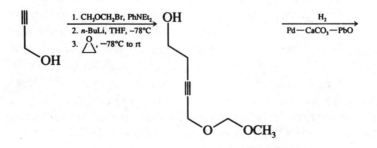

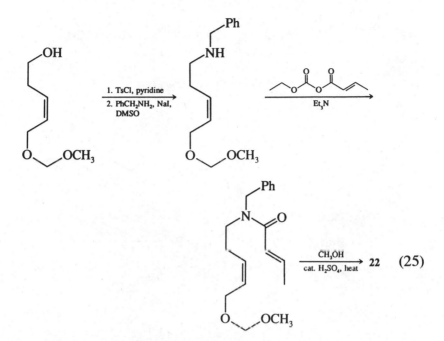

Lactone **23** was constructed by using acetylide chemistry at two sites. The disconnections follow the principles given earlier, that is, open the lactone, disconnect beside the alcohol, and so forth (Scheme XVI). The synthesis devised by Jakubowski and co-workers is shown in Eq. 26.[34]

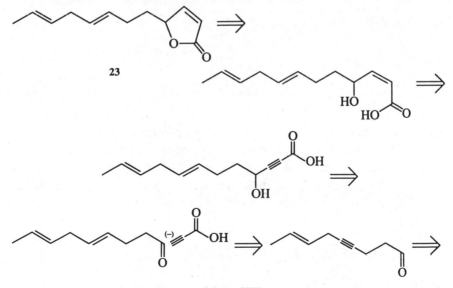

**Scheme XVI**

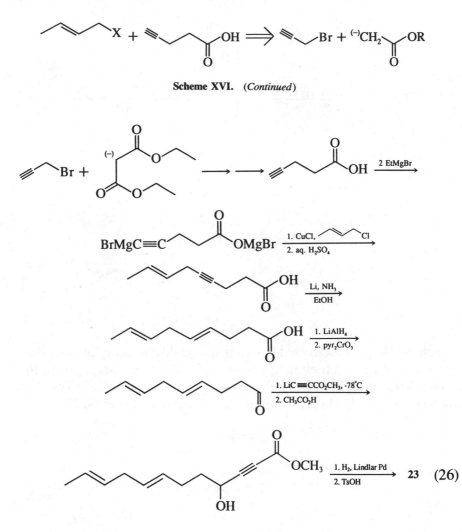

**Scheme XVI.** (*Continued*)

Two equivalents of ethylmagnesium bromide were used in order to de-protonate the carboxylic acid and the acetylide. The reactivity of di-anions is generally greater at the last site of proton removal. Lithium in liquid ammonia with ethanol gave the trans alkene, and several steps later, hydrogenation was used to prepare the cis double bond for ring closure.

## 6.6  THE DIELS–ALDER REACTION

In the Diels–Alder reaction, a diene and a dienophile combine through a cyclic transition state to give a six-membered ring. Good reactivity is found when the reacting double bond of the dienophile is electron-poor owing to conjugation with one or more electron-withdrawing groups such as esters or nitriles. On the other hand, the diene is more reactive with electron-donating groups attached. Thus in planning a synthesis, one should select an appropriately conjugated dienophile even when this may require a subsequent reduction of a carbonyl group. Examples include propenoates, propynoates, maleates, and dimethyl acetylenedicarboxylate. Other dienophiles are reactive because they contain ring strain that is partly relieved on reacting. This factor contributes to the reactivity of cyclopropene, cyclobutadiene, and cyclopentadiene. The very unstable transient benzyne is also a reactive dienophile. There are other Diels–Alder combinations in which the diene is electron-poor and the dienophile is electron-rich. These less common cases are called *reverse-electron-demand reactions* (Chapter 5, Eq. 37).

The product of a Diels–Alder reaction is generally a cyclohexene; thus, finding that feature in a structure suggests that disconnection (Scheme XVII). The cyclohexene may be part of a bicyclic or fused-

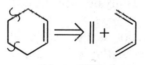

**Scheme XVII**

ring structure, or it may be a cyclohexadiene as when an activated acetylene is the dienophile. Furthermore, if the ring is hydrogenated, aromatized, or modified with new functionality, the simple presence of a six-membered ring may be sufficient reason to propose a Diels–Alder reaction step.

Compound **24**[35] is obviously a cyclohexene. Disconnecting it according to Scheme XVII gives *trans*-1,3-pentadiene and an $\alpha,\beta$-unsaturated ester (Scheme XVIII). The latter is a reactive dienophile, and it is the

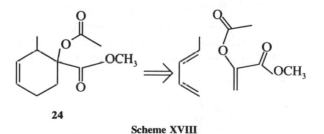

**24**

**Scheme XVIII**

enol ester of methyl pyruvate, made with acetic anhydride and TsOH. The dienophile and diene in this example are both unsymmetric; and therefore, a reversed relative orientation could give a different regio-isomer. When the diene and dienophile each have one substituted site, the major product is generally that with the substituents arranged 1,4 or 3,4 on the cyclohexene. Therefore, **24** should be the major product. When Lewis acid catalysis is used (Section 5.3), regioselectivity is even greater.

In compound **25** the cyclohexene is part of a bicyclic structure; there-fore, disconnection gives a monocyclic diene (Scheme XIX). With die-

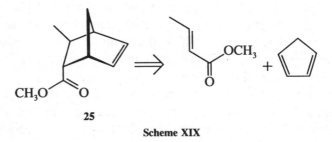

25

Scheme XIX

thylaluminum chloride catalysis, this reaction gave racemic **25** in 84% yield.[36]

The Diels–Alder reaction is stereospecific. The diene and dienophile approach with one face of the $\pi$ bonding of each merging to form $\sigma$ bonds, while the original geometry is minimally shifted (Section 8.3.2). This is a syn addition on both components. Because of this, groups that are cis in the dienophile remain cis in the cyclohexene, and the groups that are cis,cis (or trans,trans) in the diene become cis in the cyclo-hexene. Although four stereogenic centers are formed in many Diels–Alder reactions, often only one or two pairs of enantiomers are formed in appreciable amounts. Of the maximum of 32 regio- and stereoisomers imaginable with four stereogenic centers, the stereochemistry of the diene and dienophile limit it to a maximum of 8. Since both faces of cyclopentadiene are identical (Section 3.8), there is no regiochemistry in the formation of **25**, further reducing the possibilities to 4 stereoiso-mers (**25–28**). There is generally a favoring among diastereomers toward those with the final $\pi$ bonds facing each other as in **25** and **26,** where the carbonyl and alkene are close (endo rule). In this case the ratio of (**25** + **26**) to (**27** + **28**) was 15:1.

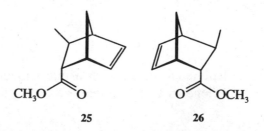

25                    26

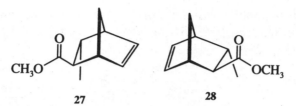

27                              28

Narrowing Diels–Alder reaction products to one enantiomer requires a chiral influence such as a template (Section 3.5).[37] A dienophile containing enantiopure 1-menthol gave only one stereoisomer in reactions with various dienes (Eq. 27).[38] Removal of the menthol template gave

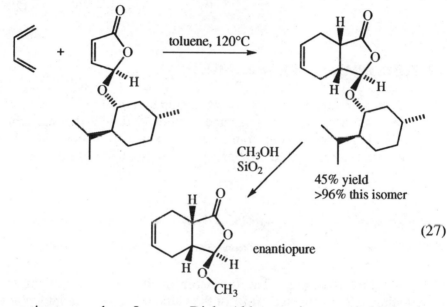

toluene, 120°C

45% yield
>96% this isomer

CH₃OH
SiO₂

enantiopure

(27)

enantiopure product. In many Diels–Alder reactions, steric hindrance or intramolecular restrictions will limit the number of isomers. Altogether, stereospecificity, regiospecificity, and the endo selectivity make most Diels–Alder reactions quite practical.

Compound **29** is a simple case with no stereochemistry, but it is not a cyclohexene. The enol tautomer would be a cyclohexene, and with this idea we can disconnect as in Scheme XX. Allyl alcohol is not

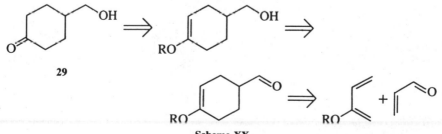

29

**Scheme XX**

reactive as a dienophile, but acrolein has the activating carbonyl group, which can be reduced later. The silyl enol ether of 1-buten-3-one is a good diene (Eq. 28).[39]

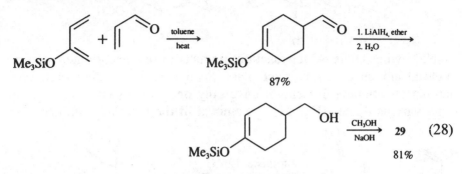

87%

29    (28)

81%

## 6.7 THE CLAISEN REARRANGEMENT

The Claisen rearrangement is used for the preparation of $\gamma,\delta$-unsaturated aldehydes, ketones, acids, esters, and amides.[40] It is a thermal rearrangement of an ether derived from an enol and an allyl alcohol (Scheme XXI). In effect, a rearranged allyl group becomes attached to the carbon

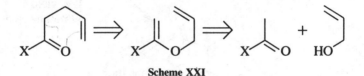

Scheme XXI

alpha to a carbonyl group. The formation of the enol ether requires a dehydrating reagent or a derivative of the carbonyl compound into which the allyl alcohol can be exchanged. These include another enol ether, an ortho ester, or an amide acetal. Examples are shown in Eqs. 29–32.[41–44]

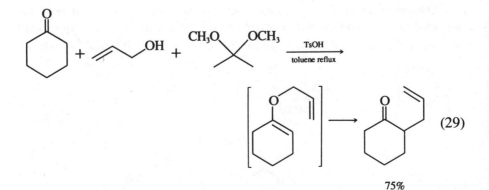

(29)

75%

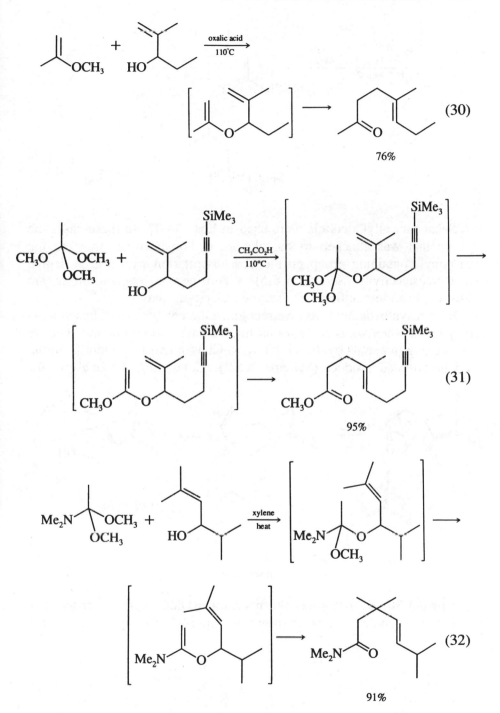

One may also begin with an allyl ester and prepare the silyl enol derivative for rearrangement (Eq. 33).[15]

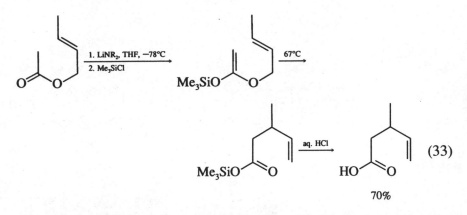

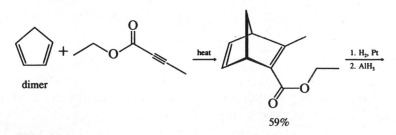

(33)

70%

Secondary allyl alcohols were used in Eqs. 30–32. In these cases the group that was attached to the alcohol carbon becomes trans to the carbonyl containing group on the $\beta,\gamma$ unsaturation with a high degree of stereoselectivity (Section 3.6.5). Where new stereogenic atoms are formed, enantiospecificity or diastereoselectivity may occur.[46]

In retrosynthetic analysis, we recognize the need for the Claisen rearrangement when we see a $\gamma,\delta$-unsaturated carbonyl compound. Devise the starting materials by drawing a retro-Claisen rearrangement from the $\gamma,\delta$-unsaturated product (Scheme XXII). Continuing the analysis for

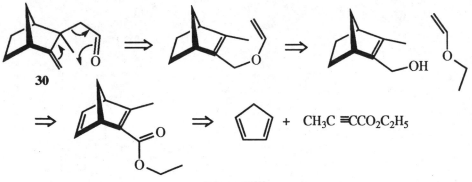

**30**

$\Rightarrow$    +    $CH_3C \equiv CCO_2C_2H_5$

**Scheme XXII**

compound **30,** we recognize the need for a Diels–Alder reaction also. The actual synthesis of **30** is shown in Eq. 34.[47] Aluminum hydride is

dimer

heat

1. $H_2$, Pt
2. $AlH_3$

59%

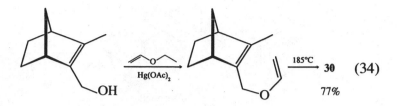

$$185°C \quad \textbf{30} \quad (34)$$

77%

used to reduce some $\alpha,\beta$-unsaturated esters where LiAlH$_4$ tends to reduce the $\alpha,\beta$-unsaturation.

Compound **31** contains unsaturation $\gamma,\delta$ to the bromo-functional carbon. Propose a carbonyl compound as a likely intermediate and then
write a retro-Claisen rearrangement (Scheme XXIII). The ketene acetal

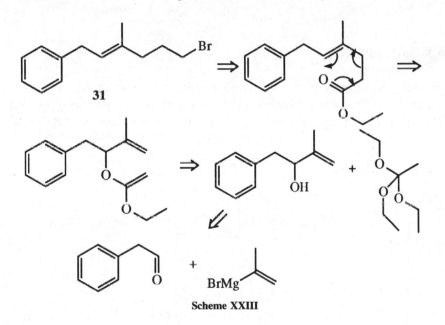

**Scheme XXIII**

could come from ethyl orthoacetate. Continue with a disconnection at
the alcohol.

The actual synthesis is shown in Eq. 35.[48]

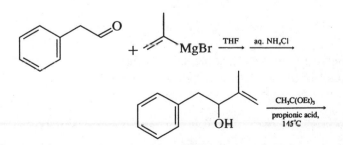

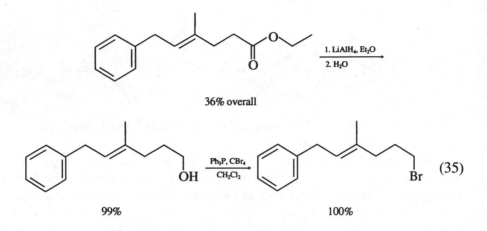

36% overall

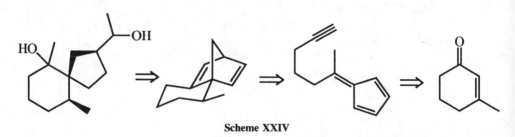

99%                                          100%

## 6.8   FINAL NOTE

In longer syntheses it is not always routine to apply certain key steps. There are highly novel creations that few persons would bring together. To pick one example among a great many, would Scheme XXIV seem

**Scheme XXIV**

obvious?[49] On the other hand, bringing together reactions for a complex scheme from a list larger than a person could bring to mind can now be done by computer. Programs have been written, backed by large data collections, that use retrosynthetic analysis to provide reaction schemes for the synthesis of complex molecules.[50]

## PROBLEMS

Show how you would synthesize each of the following compounds from simple, readily available materials.

**1.** 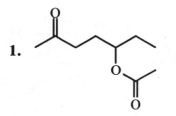 Ref. 51

**2.** 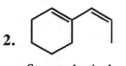 Ref. 52

Stereoselectively

**3.**  Ref. 53

**4.** 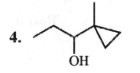 Ref. 54

**5.** 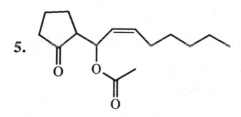 Ref. 55

**6.** 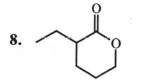 Ref. 56

**7.** 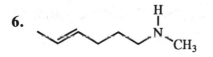 Ref. 57

**8.** 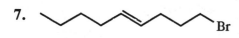 Ref. 58

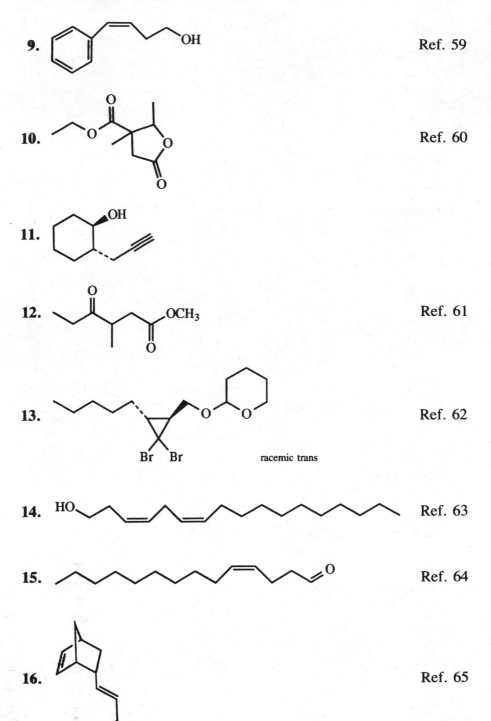

9.                                                    Ref. 59

10.                                                   Ref. 60

11.

12.                                                   Ref. 61

13.                          racemic trans           Ref. 62

14.                                                   Ref. 63

15.                                                   Ref. 64

16.                                                   Ref. 65

**17.**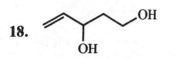

racemic cis

Ref. 66

**18.**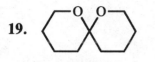

Ref. 67

**19.**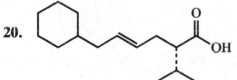

Ref. 68

**20.**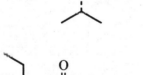

Ref. 69

**21.**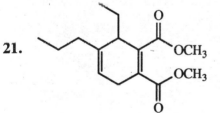

Ref. 70

**22.**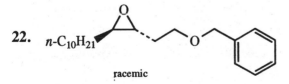

racemic

Ref. 71

**23.**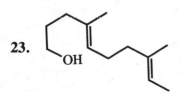

Ref. 72

**24.** 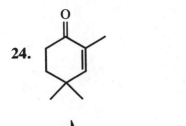                                   Ref. 73

**25.** 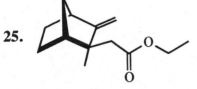                                   Ref. 74

**26.** 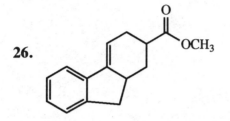                                   Ref. 75

**27.** 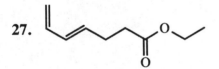                                   Ref. 76

**28.** 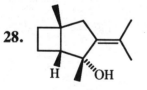                                   Ref. 77

**29.** 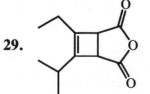                                   Ref. 78

**30.** 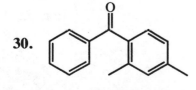                                   Ref. 79

**31.**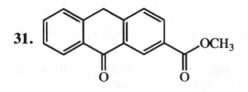

**32.** 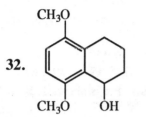                                    Ref. 80

**33.** 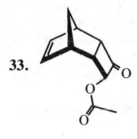                                    Ref. 81

**34.** 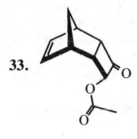                                    Ref. 82

## REFERENCES

1. Warren, S. *Organic Synthesis: The Disconnection Approach*, Wiley, New York, 1982.
2. Ellison, R. A. *Synthesis* **1973,** 397; Ho, T. L. *Synth. Commun.* **1974,** 265; and many other papers.
3. Evans, D. A.; Andrews, G. C. *Acc. Chem. Res.* **1974,** *7*, 147.
4. Harding, K. E.; Burks, S. R. *J. Org. Chem.* **1984,** *49*, 40.
5. Yamachita, M.; Matsumiya, K.; Tanabe, M.; Suemitsu, R. *Bull. Chem. Soc. Jpn.* **1985,** *58*, 407.
6. Bertz, S. H.; Dabbagh, G. *J. Org. Chem.* **1984,** *49*, 1119.
7. Seebach, D. *Angew. Chem. Internatl. Ed.* **1979,** *18*, 239.
8. Lever, Jr., O. W. *Tetrahedron* **1976,** *32*, 1943.
9. Richman, J. E.; Herrman, J. L.; Schlessinger, R. H. *Tetrahedron Lett.* **1973,** 3267.

10. Stowell, J. C.; King, B. K. *Synthesis* **1984,** 278.

11. Bellas, T. E.; Brownlee, R. G.; Silverstein, R. M. *Tetrahedron* **1969,** *25*, 5149.

12. McCormick, J. P.; Shinmyozu, T.; Pachlatko, J. P.; Schapr, T. R.; Gardner, J. W.; Stipanovic, R. D. *J. Org. Chem.* **1984,** *49*, 34.

13. Ruppert, J. F.; White, J. D. *J. Am. Chem. Soc.* **1981,** *103*, 1808.

14. Yamagiwa, S.; Kosugi, H.; Uda, H. *Bull. Chem. Soc. Jpn.* **1978,** *51*, 3011.

15. McMurry, J. E.; Melton, J. *J. Org. Chem.* **1973,** *38*, 4367.

16. Welch, S. C.; Chayabunjonglerd, S. *J. Am. Chem. Soc.* **1979,** *101*, 6768.

17. Zoretic, P. A.; Ferrari, J. L.; Bhakta, C.; Barcelos, F.; Branchard, B. *J. Org. Chem.* **1982,** *47*, 1327.

18. Kuehne, M. E.; Kirkeno, C. L.; Matsko, T. H.; Bohnert, J. C. *J. Org. Chem.* **1980,** *45*, 3259; Stork, G.; Brizzollara, A.; Landesman, H. K.; Smuszkovicz, J.; Terrel, R. *J. Am. Chem. Soc.* **1963,** *85*, 207.

19. White, W. L.; Anzeveno, P. B.; Johnson, T. *J. Org. Chem.* **1982,** *47*, 2379.

20. Kane, V. V.; Doyle, D. L.; Ostrowski, P. C. *Tetrahedron Lett.* **1980,** *21*, 2643.

21. Plesnicar, B. *Oxidation in Organic Chemistry*, W. S. Trahanovsky, Ed., Vol. **5-C,** Academic Press, New York, p. 254.

22. Stowell, J. C. *J. Org. Chem.* **1976;** *41*, 560.

23. Gottschalk, F. J.; Weyerstahl, P. *Chem. Ber.* **1975,** *108*, 2799.

24. Bloomfield, J. J.; Owsley, D. C.; Nelke, J. M. *Org. React.* **1976,** *23*, 259.

25. Hart, H.; Curtis, O. E., Jr. *J. Am. Chem. Soc.* **1956,** *78*, 112.

26. Kumar, V. T. R.; Swaminathan, S.; Rajagopalan, K. *J. Org. Chem.* **1985,** *50*, 5867.

27. Bloomfield, J. J. *Tetrahedron Lett.* **1968,** 587.

28. Bauslaugh, P. G. *Synthesis* **1970,** 287–300.

29. Ikeda, M.; Uno, T.; Homma, K.; Ohno, K.; Tamura, Y. *Synth. Commun.* **1980,** *10*, 438.

30. White, J. D.; Matsui, T.; Thomas, J. A. *J. Org. Chem.* **1981,** *46*, 3377.

31. Cargill, R. L.; Wright, B. W. *J. Org. Chem.* **1975,** *40*, 120.

32. Pak, C. S.; Kim, S. K. *J. Org. Chem.* **1990,** *55*, 1954.

33. Martin, S. F.; Bemage, B. *Tetrahedron Lett.* **1984,** *25*, 4863.

34. Jakubowski, A. A.; Guziec, F. S., Jr.; Sugiura, M.; Tam, C. C.; Tishler, M.; Omura, S. *J. Org. Chem.* **1982,** *47*, 1221.

35. Ireland, R. E., Courtney, L.; Fitzimmons, B. J. *J. Org. Chem.* **1983,** *48*, 5186.

36. Callant, P.; Storme, P.; Van der Eycken, E.; Vanderwalle, M. *Tetrahedron Lett.* **1983**, *24*, 5797.

37. Oppolzer, W. *Angew. Chem. Internatl. Ed.* **1984**, *23*, 876–889.

38. Feringa, B. L.; de Jong, J. C. *J. Org. Chem.* **1988**, *53*, 1125.

39. Yin, F.-K.; Lee, J.-G.; Borden, W. F. *J. Org. Chem.* **1985**, *50*, 531.

40. Reviews: Bennett, G. B. *Synthesis* **1977**, 589; Rhoads, S. J.; Raulins, N. R. *Org. React.* **1975**, *22*, 1.

41. Crandall, J. K.; Magaha, H. S.; Henderson, M. A.; Widener, R. K.; Thark, G. A. *J. Org. Chem.* **1982**, *47*, 5372.

42. Faulkner, D. J.; Petersen, M. R. *Tetrahedron Lett.* **1969**, 3243.

43. Johnson, W. S.; Yarnell, T. M.; Myers, R. F.; Morton, D. R.; Boots, S. G. *J. Org. Chem.* **1980**, *45*, 1254.

44. Majewski, M.; Snieckus, V. *Tetrahedron Lett.* **1982**, *23*, 1343.

45. Ireland, R. E.; Mueller, R. H. *J. Am. Chem. Soc.* **1972**, *94*, 5897.

46. Ireland, R. E.; Wipf, P.; Armstrong, III, J. D. *J. Org. Chem.* **1991**, *56*, 650.

47. Gibson, T.; Barnes, Z. J. *Tetrahedron Lett.* **1972**, 2207.

48. Guthrie, A. E.; Semple, J. E.; Joulie, M. M. *J. Org. Chem.* **1982**, *47*, 2369.

49. Nyström, J.-E.; Helquist, P. *J. Org. Chem.* **1989**, *54*, 4696.

50. A complete journal issue including eleven articles is dedicated to computer assisted synthesis: *Rec. Trav. Chim.* **1992**, *111*, 239–334.

51. Johnston, B. D.; Oehlschlager, A. C. *J. Org. Chem.* **1986**, *51*, 760.

52. Inman, W. D.; Sanchez, K. A. J.; Chaidez, M. A.; Paulson, D. R. *J. Org. Chem.* **1989**, *54*, 4872.

53. Watanabe, Y.; Iida, H.; Kibayashi, C. *J. Org. Chem.* **1989**, *54*, 4090.

54. Bartlett, P. A.; Mori, I.; Bose, J. A. *J. Org. Chem.* **1989**, *54*, 3236.

55. Iwasaki, G.; Sano, M.; Sodeoka, M.; Yoshida, K.; Shibasaki, M. *J. Org. Chem.* **1988**, *53*, 4864.

56. Bougeois, J.-L.; Stella, L.; Surzur, J.-M. *Tetrahedron Lett.* **1981**, *22*, 61.

57. Hammoud, A.; Descoins, C. *Bull. Soc. Chim. Fr.* **1978**, 299.

58. Edwards, M. P.; Ley, S. V.; Lister, S. G.; Palmer, B. D.; Williams, D. J. *J. Org. Chem.* **1984**, *49*, 3503.

59. Tai, A.; Matsumura, F.; Coppel, H. C. *J. Insect Physiol.* **1971**, *17*, 181.

60. Bjorkquist, D. W.; Bush, R. D.; Ezra, F. S.; Keough, T. *J. Org. Chem.* **1986**, *51*, 3192.

61. Herrmann, J. L.; Richman, J. E.; Schlessinger, R. H. *Tetrahedron Lett.* **1973**, 3271, 3275.

62. Porter, N. A.; Ziegler, C. B., Jr.; Khouri, F. F.; Roberts, D. H. *J. Org. Chem.* **1985**, *50*, 2252.

63. Jain, S. C.; Dussourd, D. E.; Conner, W. E.; Eisner, T.; Guerrero, A.; Meinwald, J. *J. Org. Chem.* **1983,** *48*, 2266.

64. Chattopadhyay, A.; Mamdapur, V. R. *Synth. Commun.* **1990,** *20*, 2225.

65. Polniaszek, R. P.; Stevens, R. V. *J. Org. Chem.* **1986,** *51*, 3023.

66. Mulholland, R. L., Jr.; Chamberlin, A. R. *J. Org. Chem.* **1988,** *53*, 1082.

67. Reitz, A. B.; Nortey, S. O.; Maryanoff, B. E.; Liotta, D.; Monahan, III, R. *J. Org. Chem.* **1987,** *52*, 4191.

68. Reddy, G. B.; Mitra, R. B. *Synth. Commun.* **1986,** *16*, 1723.

69. Herold, P.; Duthaler, R.; Rihs, G.; Angst, C. *J. Org. Chem.* **1989,** *54*, 1178.

70. Paquette, L. A.; Melega, W. P.; Kramer, J. D. *Tetrahedron Lett.* **1976,** 4033.

71. Flippin, L. A.; Jalai-Araghi, K.; Brown, P. A.; Burmeister, H. R.; Vesonder, R. F. *J. Org. Chem.* **1989,** *54*, 3006.

72. Ho, N.-h.; le Noble, W. J. *J. Org. Chem.* **1989,** *54*, 2018.

73. Schuster, D. I.; Rao, J. M. *J. Org. Chem.* **1981,** *46*, 1515.

74. Monti, H.; Corriol, C.; Bertrand, M. *Tetrahedron Lett.* **1982,** *23*, 947.

75. Ranu, B. C.; Sarhar, M.; Chakraborti, P. C.; Ghatah, U. R. *J. Chem. Soc. Perkin Trans. I* **1982,** 865.

76. Parker, K. A.; Iqbal, T. *J. Org. Chem.* **1982,** *47*, 337.

77. Cargill, R. L.; Wright, B. W. *J. Org. Chem.* **1975,** *40*, 120.

78. Smith, A. B., III; Boschelli, D. *J. Org. Chem.* **1983,** *48*, 1217.

79. Utz, C. G.; Shechter, H. *J. Org. Chem.* **1985,** *50*, 5705.

80. Chen, C.-P.; Swenton, J. S. *J. Org. Chem.* **1985,** *50*, 4569.

81. Miller, R. D.; Dolce, D. L.; Merritt, V. Y. *J. Org. Chem.* **1976,** *41*, 1221.

82. Angus, R. O., Jr.; Johnson, R. P. *J. Org. Chem.* **1983,** *48*, 273.

# 7

# MECHANISMS AND PREDICTIONS

When planning a new reaction in organic chemistry, we look at the accumulated information on similar reactions in order to predict the best conditions for it. The more we know about the intimate details of the reaction process at the molecular level, the better will be our predictions. A particular reaction may be described as an ordered sequence of bond breaking and making and a series of structures that exist along the way from starting material to product. The description includes the concurrent changes in potential energy. Structures at energetic maxima are called *transition states*, and structures at minima are called *intermediates*. The complete description is called the *mechanism* of the reaction.

## 7.1 REACTION COORDINATE DIAGRAMS AND MECHANISMS

The energy–structure relationship is sometimes illustrated with a plot of potential energy versus progress along the pathway of lowest maximum potential energy. This is exemplified in Fig. I for the chain propagation steps of the familiar chlorination of methane. The first maximum is the transition state at which the C—H bond is partially broken and the H—Cl bond is partially formed. At the shallow minimum, the transient intermediate methyl radical exists. The second maximum is the transition state at which a Cl—Cl bond is partially broken and a C—Cl bond is partially formed. Collisions interconvert kinetic and potential energy; thus the increase in potential energy required to reach a transition state corresponds to a decrease in kinetic energy, and a descent from a transition state corresponds to an increase in kinetic energy. Likewise, the net overall descent for this reaction corresponds to a net increase in kinetic energy; that is, the process is exothermal.

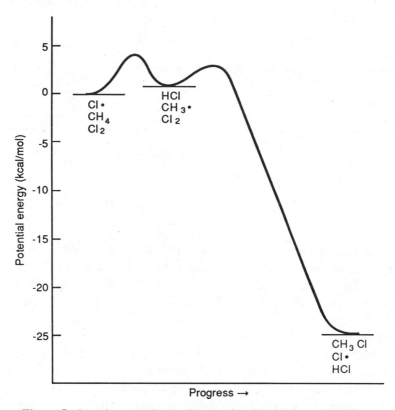

**Figure I.** Reaction coordinate diagram for chlorination of methane.

A reaction that requires a higher rise to a transition state (activation energy) will be slower than one requiring a lesser rise (if the probability factors are similar) because a smaller fraction of collisions will provide sufficient potential energy to make it.

The energy values for such plots are derived from measurements of overall exo- or endothermicity and from measurement of the effect of varying the temperature on the rate of the reaction (Section 7.3.6).

Many techniques have been developed for determining mechanisms including complete product (and sometimes intermediate) identification, isotope labeling, stereochemistry, and kinetics,[1,2] as are covered in Section 7.3. In actuality, two or more alternative mechanisms are proposed, and their differences are probed with these techniques. An observation is made that is incompatible with one of the proposed mechanisms, and that mechanism is eliminated from consideration, hopefully leaving one. Reasonable mechanisms that have withstood various experimental tests gain some acceptance and are then very useful in predicting the possible

range of applications of the reaction and for suggesting changes in reaction conditions that will improve the yield and efficiency of the reaction.

Since molecules are individually too small and too fast for direct
observation, our pictorial mechanisms are not the final word. Published
mechanisms vary from unsupported conjecture to highly tested near certainty, and you should maintain a healthy skepticism in order to improve
on what is available.

Even where the goal, a well-defined singular mechanism, is not yet
attained for a given reaction, the observations can be used to make
predictions. For example, using the Hammett equation (Section 7.3.7)
we can predict from measured rates of some cases of a reaction how
fast a new case with different substitution will occur.

## 7.2  THE HAMMOND POSTULATE

In a reaction coordinate diagram it is obvious that the potential energy
content at a transition state is closer to that in the starting materials in
an exothermal step, and closer to the products in an endothermal step.
Since potential energy is required to distort a molecule, the structure of
the transition state will more closely resemble those molecules to which
it is closer in potential energy; that is, a small vertical difference in a
reaction coordinate diagram corresponds to a small horizontal difference.
Transition states are late in endothermal steps, and early in exothermal
steps. This is the Hammond postulate,[3] and it is useful for predicting
products where there is potentially close competition between two alternative steps.

In Fig. II we see a choice of two late transition states that are nearly
as different in energy as the products are. We can predict that the thermodynamically more stable product will greatly predominate among the
products (if the probability factors are similar). In Fig. III we see a
choice of two early transition states, both resembling the starting material and thus resembling each other in structure and energy. We can
predict that there may be little or no selectivity toward the more stable
product.

Comparisons can be made for less extreme cases as well. Comparing
two endothermal reactions, we can predict that the more endothermal
one will be the more selective. For example, hydrogen atom abstraction
from propane by Br atoms is more endothermal and more selective than
by chlorine atoms.

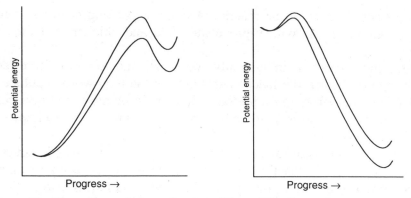

**Figure II.** Competing endothermal steps.   **Figure III.** Competing exothermal steps.


## 7.3   METHODS FOR DETERMINING MECHANISMS

### 7.3.1   Identification of Products and Intermediates

The primary information should be a thorough identification of the products of the reaction under investigation. If a mechanism proposal includes a temporary intermediate compound that may have some stability, attempts should be made to isolate some of it from the reaction by premature interruption. If it is not stable enough for this, it may yet be detectable spectroscopically in the reaction mixture while in progress. It may also be possible to divert the intermediate by adding a new reactant to the mixture. Finally, if the intermediate is a stable compound and can be prepared by another means, it can be used as a substitute starting material to determine whether it does give the same products overall, at least as rapidly.


### 7.3.2   Isotope Tracing

If competing mechanistic proposals differ in what atom in the starting material becomes a particular atom of the product, a determination may be made using isotope labeling. A starting material may be synthesized with an uncommon isotope in a specific site in the molecule. A reaction may then be carried out using this labeled material, and the product then examined for the presence or absence of the uncommon isotope, or to determine the site in the product wherein the isotope resides. Isotopes available for tracing the origin of atoms in the product include $^2H$, $^3H$, $^{13}C$, $^{15}N$, and $^{18}O$. The presence of the radioactive isotopes in the products or degradation derivatives of the products is determined by decay counting. Mass spectra will show the presence and sometimes location of isotopes if the product is compared with the unlabeled compound. Peaks will be displaced toward higher mass for the fragment ions or molecular ions where heavier isotopes are present. NMR spectroscopy is especially useful for isotope tracing. If a deuterium is present in a product, the signal for that site will be missing or smaller in the $^1$H NMR spectrum of the product. The deuterium atoms themselves may be detected at their resonance frequency. The low natural abundance of $^{13}$C gives weak $^{13}$C spectra, but if a compound is enriched in that isotope at a particular site, the signal for that site will be obviously stronger in the spectrum.

Carbon-13 NMR was used to determine which group migrated in the reaction shown in Eq. 1.[4] Treatment of the ortho-substituted biphenyl derivative with $AlCl_3$ gave the corresponding meta compound. It is not obvious whether the phenyl group migrated or the aminomethyl group migrated. This was determined by synthesizing a sample of the ortho compound with a greater than natural amount of $^{13}$C at the * site shown in Eq. 1. The $^{13}$C NMR spectrum (without proton decoupling) of the

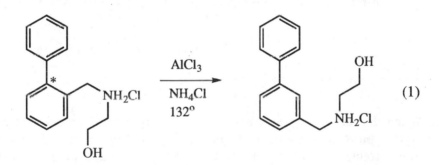

$$(1)$$

labeled starting material showed the expected weak signals plus a strong singlet at 142.36 ppm for the labeled carbon site. The product of the reaction showed weak signals plus a strong doublet at 129.07 ppm, $J_{CH}$ = 158.2 Hz. The starting singlet indicated that there was no hydrogen directly attached to the labeled site, as expected, but the product doublet indicated that the labeled carbon now has a hydrogen attached. One may conclude with confidence that the phenyl group migrated. ($^{13}$C NMR spectroscopy is summarized in Chapter 10.)

### 7.3.3 Stereochemical Determination

If a reaction is carried out on a particular stereoisomer of starting material and the products are stereoisomerically identified (Chapter 3), a choice among mechanisms can often be made. In substitution and rearrangement reactions, inversion of configuration will indicate back-side attack, retention will indicate front-side attack, and racemization will indicate formation of a flattened (achiral) intermediate such as a carbocation. If two steps have occurred between starting material and product, the interpretations will differ. Retention could be the result of two inversions, and racemization could be the result of some inverting attack by the former leaving group.

One may determine whether additions to alkenes are syn, anti, or mixed. For example, trifluoromethyl hypochlorite adds to alkenes, and one may propose the following mechanisms:

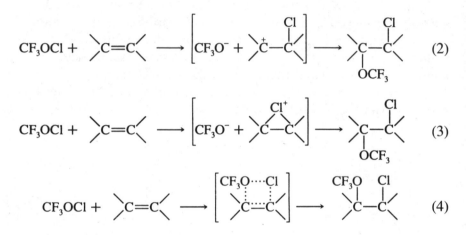

When this addition was carried out with pure *cis*-2-butene, only the erythro product was obtained, and with pure *trans*-2-butene, only the threo product was obtained.[5] If Eq. 2 were the mechanism, a mixture would be expected; if Eq. 3 were the mechanism, the results would have been the opposite. Only Eq. 4 is in accord with the stereochemical results, that is, a concerted (or nearly so) syn addition.

In elimination reactions, a similar comparison of the stereochemistry of starting materials and products can indicate syn, anti, or mixed processes. One method for changing the stereochemistry of a double bond includes epoxidation and ring opening with lithium diphenylphosphide, followed by oxidation and elimination (Eq. 5).[6] The epoxide from *cis*-

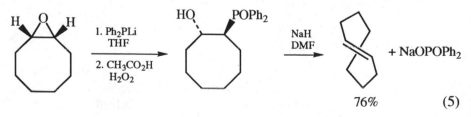

clyclooctene gives the *trans*-hydroxy phosphine intermediate. Since the final product is *trans*-cyclooctene, the deprotonated hydroxy and the phosphine oxide groups must be oriented side-by-side for P—O bonding while they depart as phosphinate. Thus the elimination is syn.

### 7.3.4  Concentration Dependence of Kinetics

The measurement of rates of reactions under various conditions gives several different kinds of mechanistic information.[7] The concentration dependence of rates can give information on the number of steps in a mechanism and the species involved in reaction collisions. For these purposes other variables such as temperature and choice of solvent are kept constant.

For reactions in homogeneous solution, the concentrations of components are measured as time progresses; the usual units are moles of solute per liter of solution. At any one moment during the progress of a reaction, a certain number of moles of product are appearing per liter of solution per second. This is called the *rate of the reaction*, and it generally decreases as the supply of starting material is depleted. This is shown graphically in Fig. IV for the simple reaction in Eq. 6. The rate at any one moment is the slope of the graph of [B].

$$A \rightarrow B + C \tag{6}$$

At the same time the concentration of A is decreasing; thus the slope of the curve for [A] is equal but opposite in sign to that for [B]. There-

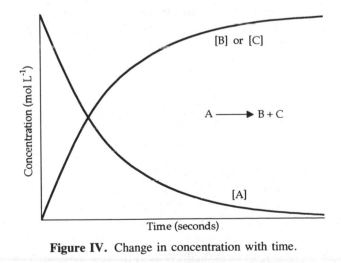

**Figure IV.** Change in concentration with time.

fore, two alternative rate expressions may be written:

$$\text{Rate} = \frac{d[B]}{dt} \quad \text{or} \quad \text{rate} = \frac{-d[A]}{dt} \tag{7}$$

(where rate is in units of moles per liter per second).

Experimentally, a solution of known concentration of starting material A is prepared, and then as the reaction proceeds, the concentration of A and/or B is measured repeatedly. In this simple reaction, A continually decomposes without coreactants or catalysts. When the supply of A reaches half the original concentration, the rate should be half the initial rate. If the rate at any point in time is divided by [A] at that point, the quotient should be a constant $k$, sometimes called the *rate constant* or *specific rate* (Eq. 8). The constant $k$ is simply the rate of the reaction when $[A] = 1$ (even though it may have been determined at much lower concentrations than 1 molar).

$$\frac{\text{Rate}}{[A]} = k \quad \text{or} \quad \frac{-d[A]}{dt} = k[A] \tag{8}$$

(where $k$ is expressed in reciprocal seconds, $\text{sec}^{-1}$).

The determined value of $k$ is a fundamental property of the reaction, independent of the concentration of A. Since by Eq. 7 the rate is proportional only to the concentration of one component to the first power, this is called a *first-order reaction*. A simple first-order reaction mechanism involves only collisions of reactant with unreactive solvent molecules that provide the kinetic energy necessary for bond reorganization. The change in $k$ with a change in temperature or structure of A gives other mechanistic information, as is shown in Sections 7.3.6 and 7.3.7.

Measuring the slopes along a curve in Fig. IV is not an efficient way to obtain $k$. It is more readily extracted from a linear plot that results from the integrated form of Eq. 8:

$$-\int_{[A]_0}^{[A]_t} \frac{d[A]}{[A]} = k \int_0^t dt; \quad \ln \frac{[A]_0}{[A]_t} = kt \tag{9}$$

Plotting the natural logarithm of the ratio of an initial [A] to the [A] at time $t$ versus time gives a straight line of slope $k$ (Fig. V). The value of $k$ from this linear plot is based on the whole set of points graphically or least-squares-optimized. Notice also that the actual value of [A] does not need to be measured; only a ratio of values of [A] is needed. This sort of ratio is easily obtained spectroscopically as ratios of areas of

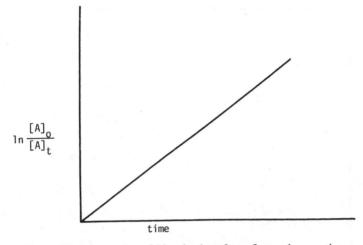

**Figure V.** Linear plot of kinetic data for a first-order reaction.

diminishing peaks in a series of NMR spectra or ratios of absorbance from a spectrophotometer. Finally, if the data do not give a straight line in such a plot, the reaction is not first-order.

Dimerization reactions (Eq. 10) give a different kinetic result. Here a collision between two A molecules is necessary; therefore, as the concentration of A decreases, the rate drops faster than found for Eq. 6.

$$A + A \rightarrow B \tag{10}$$

At half the initial concentration, collisions would be only one-fourth as frequent. Division of the rate by $[A]^2$ gives a constant for this reaction (Eq. 11). Since the rate is proportional to the concentration of A squared, it is called a *second-order reaction*.

$$\frac{Rate}{[A]^2} = k \quad \text{or} \quad \frac{-d[A]}{dt} = k[A]^2 \tag{11}$$

Once again the data give a linear plot with a slope of $k$ if the rate equation is integrated:

$$-\int_{[A]_0}^{[A]_t} \frac{d[A]}{[A]^2} = k \int_0^t dt; \quad \frac{1}{[A]_t} - \frac{1}{[A]_0} - kt \tag{12}$$

The dependence of the rate on concentration in first-, second-, and third-order reactions is displayed graphically in Fig. VI. The $y$ axis is the slope taken from graphs of the sort in Fig. IV at various $[A]$ values,

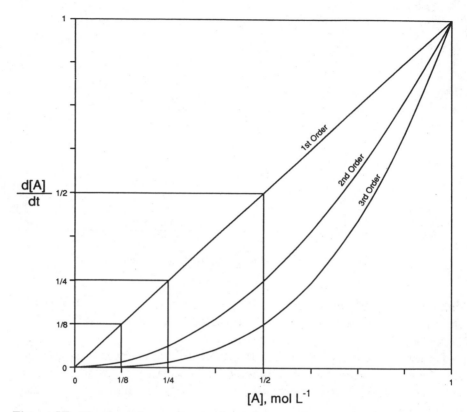

**Figure VI.** The dependence of rate of reactions on the concentration of a starting component A when first, second, and third order in A. The rates for all three are arbitrarily set at 1 when [A] is 1 molar.

and the $x$ axis is [A]. For simplicity the rate is arbitrarily set at 1 where [A] is 1 for all three reactions. In the first-order reaction you can see that the rate drops to $\frac{1}{2}$ when [A] is $\frac{1}{2}$, and the relationship is linear. In the second-order reaction, the rate drops to $\frac{1}{4}$ when [A] is $\frac{1}{2}$ and the curve is a parabola. In the third-order reaction the rate drops to $\frac{1}{8}$ when [A] is $\frac{1}{2}$, and the curve is cubic.

If the log of the rate is plotted against the log of [A], all the orders give straight lines with slopes of 1, 2, and 3, respectively (Eq. 13). This relationship allows an easy determination of the order ($n$) in A using two rate values (Eq. 14).

$$\frac{-d[A]}{dt} = k[A]^n$$

$$\log \text{rate} = \log k + n \log [A] \tag{13}$$

$$\log \text{rate}_1 - n \log [A]_1 = \log k = \log \text{rate}_2 - n \log [A]_2$$

$$\log \frac{\text{rate}_1}{\text{rate}_2} = n \log \frac{[A]_1}{[A]_2} \tag{14}$$

Substitute two rates and their corresponding [A] values into Eq. 14 and solve for $n$. If the rate depends on the concentrations of other compounds at the same time, simply keep their concentrations constant for the rate measurements to obtain the order in A.

A simple bimolecular reaction with two different starting materials A and B (Eq. 15) gives a rate equation (Eq. 16) showing that it is first-order in A and first-order in B or second-order overall.

$$A + B \rightarrow C \tag{15}$$

$$\frac{-d[A]}{dt} = k[A][B] \tag{16}$$

There are three ways to analyze data for this kind of reaction to confirm the order and determine the value of $k$:

1. One may begin with equal concentrations of A and B and use Eq. 11.

2. One may use a large excess of B so that as $[A]_t$ diminishes, the change in [B] is negligible; that is, $[B]_0 = [B]_t$. Integration of Eq. 16 assuming [B] is a constant gives Eq. 17.

$$\ln \frac{[A]_0}{[A]_t} = k[B]t \tag{17}$$

Thus plotting $\ln[A]_0/[A]_t$ versus $t$, as in Eq. 9, gives a straight line of slope $k[B]$. Dividing the slope by the known [B] gives the second-order rate constant. This use of a large excess of B is called *pseudo-first-order conditions*. Some authors tabulate $k[B]$ values labeled as $k_{obs}$ or $k_{pseudo}$. It is necessary to repeat the reaction with one or more other concentrations of B to be certain that the reaction is first-order in [B] (not zero or second-order in [B], which would still give pseudo-first-order results).

3. The integrated form of the rate equation (both [A] and [B] as variables) can be plotted (Eq. 18). It is common to plot only the variables

$\ln([A]_t/[B]_t)$ versus $t$ to check for linearity and to obtain the slope,[8] from which $k$ is calculated using Eq. 18.

$$\frac{1}{[A]_0 - [B]_0} \ln \frac{[B]_0[A]_t}{[A]_0[B]_t} = kt \tag{18}$$

Pseudo-first-order conditions can be used whatever the order in the reactant B used in large excess. The $k_{pseudo}$ values obtained at two different concentrations of B can be used to determine the order in B using Eq. 19, which is derived from $k_{pseudo} = k[B]^n$ in analogy to Eq. 14.

$$\log \frac{k_{pseudo\,1}}{k_{pseudo\,2}} = n \log \frac{[B]_1}{[B]_2} \tag{19}$$

Some reactions give kinetic orders that do not match the stoichiometry of the process. A reactant may not appear in the kinetic expression (zero-order) or another may appear with an order higher or lower than the number of equivalents consumed in the process. Some reactions may total third-order or higher. Such kinetic characteristics indicate multistep mechanisms (problems 6 and 7 at the end of the chapter). On the other hand, some multistep mechanisms may exhibit simple first- or second-order kinetics, and require other evidence to delineate the mechanism.

Consider the two-step process shown in Eq. 20, where an intermediate compound I is formed in the first step and consumed in the second.

$$
\begin{array}{c}
A + B \xrightarrow{k_1} I \\
I + C \xrightarrow{k_2} D
\end{array}
\tag{20}
$$

If the first step is relatively slow and the second step fast, then I will be consumed as rapidly as it is formed. The concentration of I will remain very low and practically constant. Assuming that constancy (known as the *steady-state approximation*), we can equate the rate of formation of I and the rate of consumption of I, which also equals the rate of formation of D:

$$\frac{d[D]}{dt} = k_1[A][B] = k_2[I][C] \tag{21}$$

Thus, in terms of measurable concentrations, the reaction is second-order just as for Eq. 15, and reactant C does not affect the rate.

In other reactions the first step may be fast and reversible, followed by a slow step:

$$A + B \underset{k_{-1}}{\overset{k_1}{\rightleftharpoons}} I$$

$$I \xrightarrow{k_2} C \qquad (22)$$

The first step provides a low concentration of I that is related by an equilibrium constant to [A] and [B] (Eq. 23).

$$K = \frac{[I]}{[A][B]} \qquad (23)$$

The rate of formation of C is proportional to [I] which from Eq. 23 is proportional to [A][B] (Eq. 24). Once again we would observe simple second-order kinetics indistinguishable from Eqs. 15 and 20. Detection of I and determination of the equilibrium constant would distinguish these.

$$\frac{d[C]}{dt} = k_2[I]$$

$$\frac{d[C]}{dt} = k_2 K[A][B] \qquad (24)$$

In the following example reaction of this sort,[9] the prior equilibrium became apparent in the curvature of a plot as will be seen below. Treatment of 1,1-diphenyltrichloroethanol with sodium hydroxide caused elimination of chloroform. The proposed mechanism is shown in Eq. 25. In this reaction the anionic intermediates were present in low con-

$$\underset{\underset{Ph_2CCCl_3}{|}}{\overset{OH}{|}} + OH^- \rightleftharpoons \underset{\underset{Ph_2CCCl_3}{|}}{\overset{:\ddot{O}:^-}{|}} + H_2O$$

$$\underset{\underset{Ph_2CCCl_3}{|}}{\overset{:\ddot{O}:^-}{|}} \longrightarrow Ph_2C{=}\ddot{O} + :\bar{C}Cl_3$$

$$:\bar{C}Cl_3 + H_2O \longrightarrow HCCl_3 + :\ddot{O}H^- \qquad (25)$$

centrations and not detected directly. A unique feature here is that the concentration of OH⁻ does not change because it is regenerated in the

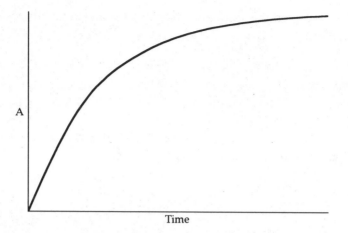

**Figure VII.** Change in absorbance with progress of the reaction.

third step. Therefore, pseudo-first-order conditions exist at any concentration of OH⁻. Measurements showed that the reaction is pseudo-first-order in the chloroalcohol and also first-order in [OH⁻].

The reaction was run in the sample cell of a spectrometer, where the temperature was controlled to ±0.1°C. The absorbance at 258 nm for benzophenone was followed and plotted against time as shown in Fig. VII. Absorbance ($A$) is proportional to concentration. Figure VIII shows a plot of $A_\infty - A_t$, which is proportional to the concentration of the chloroalcohol, assuming 100% conversion:

$$[ROH]_t = j(A_\infty - A_t) \tag{26}$$

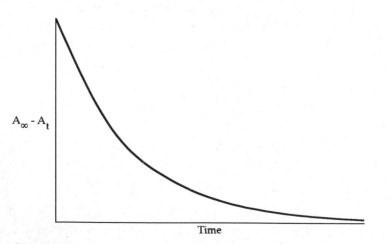

**Figure VIII.** Change in $A_\infty - A_t$ with progress of the reaction.

Rewriting Eq. 9 for this case (Eq. 27) and substituting from Eq. 26, we find Eq. 28. Since $A_0$ is zero we have Eq. 29.

$$\ln \frac{[ROH]_0}{[ROH]_t} = kt \qquad (27)$$

$$\ln \frac{j(A_\infty - A_0)}{j(A_\infty - A_t)} = kt \qquad (28)$$

$$\ln A_\infty - \ln(A_\infty - A_t) - kt \qquad (29)$$

A plot of $\ln A_\infty - \ln(A_\infty - A_t)$ versus $t$ is linear, and the slope is the pseudo-first-order rate constant $k_{obs}$. Since $\ln A_\infty$ is a constant, it sets the intercept but it does not affect the slope and may be omitted.

Numerous runs were followed, beginning with $1.5 \times 10^{-5}$ $M$ chloroalcohol and different concentrations of $OH^-$. The rate was proportional to the $[OH^-]$ in the pH range of 10–12, indicating a first-order dependence on $[OH^-]$, but at higher $[OH^-]$ the rate increase was less than proportional. This is apparent in a plot of the log of the pseudo-first-order rate constants against pH (Fig. IX). The linear portion of the plot is in accord with Eq. 25 but would just as well fit a reaction of the sort in Eq. 15. However, since the chloroalcohol has a $pK_a$ of 12, it is largely converted to alkoxide near pH 13 and additional $OH^-$ would do little to change this; therefore, the slope levels off. This is in accord

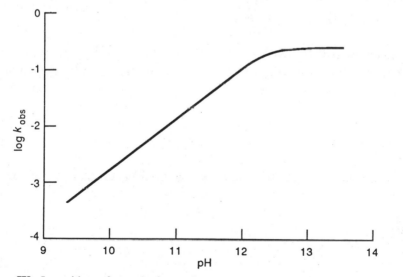

**Figure IX.** Logarithm of pseudo-first-order rate constants versus pH for the reaction of Eq. 25 at 25°C. Reprinted by permission from Nome, F.; Erbs, W.; Correia, V. R. *J. Org. Chem.* **1981**, *46*, 3802. Copyright 1981 American Chemical Society.

with the prior equilibrium in Eq. 25 but not the one-step mechanism in Eq. 15. Thus the slow step follows the preequilibrium.

A more complex expression can be written that will predict the linear and curved portions[10] but is beyond the scope of coverage here.

The mechanism in Eq. 25 is called an "E1cB" because it is an elimination (of $CHCl_3$) from the conjugate base of the chloroalcohol. The 1 signifies that the reaction is first-order in the conjugate base (step 2).

Another sort of preliminary equilibrium is illustrated in Eq. 30, where

$$A \underset{k_{-1}}{\overset{k_1}{\rightleftharpoons}} I + B$$
$$I + C \overset{k_2}{\longrightarrow} D \tag{30}$$

the starting material A dissociates rapidly and reversibly to give a low concentration of intermediate I and by-product B. The rate expression is first stated for the slow step, the consumption of I by C:

$$\frac{d[D]}{dt} = k_2[I][C] \tag{31}$$

The unmeasurable [I] is replaced using the equilibrium expression for the first step:

$$\frac{d[D]}{dt} = \frac{k_2 K[A][C]}{[B]} \tag{32}$$

This reaction will follow second-order kinetics initially, but as B accumulates, the reaction will slow down more than would be expected for Eq. 16-type behavior, since [B] is in the denominator. Obviously the equilibrium concentration of I provided by the first step will be depressed by increasing concentrations of B. In fact, the preliminary equilibrium step may be detected simply by adding extra B to the initial reaction solution and noting the slower rate.

Another common circumstance is an Eq. 30-type reaction where the first step is the slow one. As in Eq. 20, the concentration of I will remain low and practically constant and we can use the steady-state approximation:

$$k_1[A] = k_{-1}[I][B] + k_2[I][C] \tag{33}$$

The rate of formation of D depends on the [I] and [C] (Eq. 34), but substituting from Eq. 33, we find Eq. 35. Early in the reaction the

$$\frac{d[D]}{dt} = k_2[I][C] \tag{34}$$

$$\frac{d[D]}{dt} = k_1[A] - k_{-1}[I][B] \tag{35}$$

concentrations of I and B are very small, and the reaction follows first-order kinetics.

Added extra B may slow the reaction, but from the kinetics, it is obvious that there are two steps because reactant C, which is involved in the product formation, is absent from the rate expression. It must come in later in the fast step. The familiar $S_N1$ reactions are of this type.

Many other reaction sequences lead to a wide variety of rate equations, some of which are easily analyzed in ways analogous to those developed above, and many of which require more complicated treatments.[7]

### 7.3.5 Isotope Effects in Kinetics

The ground-state vibrational energy of a bond is lower for a bond to a heavier isotope than it is for a lighter isotope. Therefore, the activation energy required to break the bond with the heavier isotope will be greater than that required for the lighter isotope. The higher activation energy process is slower. The largest differences are found with hydrogen, where the mass ratios of the isotopes are greatest. For deuterium, rate constant ratios $k_H/k_D$ range up to about 10. The smaller mass ratio for $^{12}C$ to $^{13}C$ gives $k_{12C}/k_{13C}$ up to about 1.1. These ratios are called *primary kinetic isotope effects*.[11] Heavier isotopes located near a reaction site but not involved in bond breaking give a smaller secondary kinetic isotope effect.

If the rate-determining step of a reaction mechanism involves the breakage of a C—H bond, the corresponding C—D compound will be substantially slower. The maximum effect occurs if the transition state occurs at the midpoint of hydrogen transfer, and if the former bonding partner, the coming bonding partner, and the hydrogen are colinear. If the C—H bond is broken in a step subsequent to the rate-determining one, $k_H/k_D$ will be one.

A distinction was made between two proposed mechanisms for the fragmentation of an oxaziridine, using a combination of the primary kinetic isotope effect and stereochemistry.[12] Amines can behave as bases and attack at hydrogen, or they can be nucleophiles and attack at carbon or nitrogen; E2 and $S_N2$ reactions are examples. In Eq. 36 a tertiary amine (brucine) is shown in the role of base in a step resembling an E2 reaction. In Eq. 37 the amine is shown attacking the nitrogen as in an

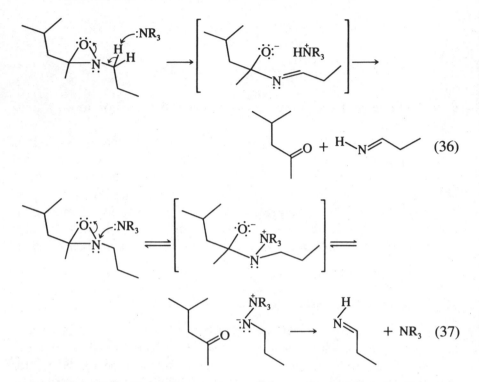

$$S_N2$$ reaction. In order to disprove one of these alternatives, the dideutero analog **1** was prepared and the rates were measured for the fragmentation of both the hydrogen and deutero compounds with various concentrations of brucine in refluxing acetonitrile. The ratio of the pseudo-first-

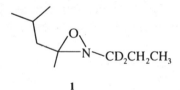

**1**

order rate constants $k_H/k_D$ was found to be 4.25. Thus the removal of the H or D is in or before the rate-determining step. This is in accord with Eq. 36, with the first step rate-determining. It would also be in accord with Eq. 37 if the last step were rate-determining. If the last step were the slow one, either intermediates would accumulate (they do not in this case) or the early steps must be fast and reversible, giving only a low concentration of intermediate. The fact that ring opening is not reversible was proved stereochemically. Inversion of configuration on nitrogen does not readily occur when the nitrogen is part of a three-membered ring; therefore, the Z and E isomers of the oxaziridine could be prepared and isolated. One isomer was treated with brucine until 43% of it was converted to products, and the remaining oxaziridine was examined by $^1$H NMR. This showed complete absence of the other diastereomer, which would have been present if some of the acyclic intermediate had reclosed. Thus Eq. 37 is not in accord with the results, and Eq. 36 remains as the most reasonable mechanism.

### 7.3.6 Temperature Effects on Kinetics

In any reaction step involving some bond breaking, there will be a potential energy high point called a *transition state* as described in Section 7.1. The higher the temperature, the larger the proportion of molecules with sufficient kinetic energy to successfully reach the transition-state condition. Determining how steeply the rate of a reaction increases with increasing temperature gives two kinds of information about the transition state: (1) the enthalpy of activation $\Delta H^{\ddagger}$, that is, the difference in $\Delta H$ (and, therefore, structure) between starting material and the activated complex at the transition state; and (2) an indication of the change in the extent of organization (entropy, $\Delta S^{\ddagger}$) of the atoms and solvent molecules as they proceed to the transition state. Reactions with a larger $\Delta H^{\ddagger}$ are accelerated to a greater extent by a temperature increase (if $\Delta S^{\ddagger}$ is similar).

The change in enthalpy and entropy from starting materials to products at 1 $M$ concentration (standard state) in a reaction may be calculated if an equilibrium constant can be measured for the reaction at several temperatures (Eqs. 38 and 39). The equilibrium constant gives the change in Gibbs free energy $\Delta G^0$ for conversion of a mole of starting materials to products all at 1 $M$. Equilibrium constants measured at two or more temperatures allow calculation of $\Delta H^0$ and $\Delta S^0$.

$$\Delta G^0 = -RT \ln K \qquad (38)$$

$$\Delta G^0 = \Delta H^0 - T \Delta S^0 \qquad (39)$$

If we could set up a simple equilibrium between starting materials and the transition state for a reaction and measure the amounts of each present at more than one temperature, we could calculate $\Delta G^0$, $\Delta H^0$, and $\Delta S^0$ for that change. However, a transition state is too short-lived, and we cannot measure the molar concentration of it to calculate the equilibrium constant for its formation.

Consider the reaction in Eq. 40 wherein starting materials A and B combine to give a short-lived transition state $AB^{\ddagger}$.

$$A + B \rightarrow [AB^{\ddagger}] \rightarrow C \tag{40}$$

Eyring's theory of absolute reaction rates enables us to determine a concentration of activated complexes from rate measurements. The activated complexes proceed on to products in the manner of a first-order reaction, and the rate constant (specific rate) is $k_B T/h$ for any reaction, where $k_B$ is Boltzmann's constant, $T$ is the absolute temperature, and $h$ is Planck's constant (Eq. 41). This is a very fast process; its lifetime is comparable to a bond vibration time. At 300 K, $k_B T/h$ is $6.3 \times 10^{12}$ $sec^{-1}$. If the molar concentration of activated complexes is $0.16 \times 10^{-12}$, the rate of formation of products is 1 mol $L^{-1}$ $sec^{-1}$. (The activated complexes are also replenished at the same rate since they are in equilibrium with starting materials.)

$$\frac{-d[A]}{dt} = \frac{k_B T}{h} [AB^{\ddagger}] \tag{41}$$

The rate of the reaction is measurable and the rate constant relating it to known concentrations of starting materials may be calculated (Eq. 42).

$$\frac{-d[A]}{dt} = k_{rate}[A][B] \tag{42}$$

Substituting from Eq. 42 into Eq. 41 we obtain Eq. 43, from which we may extract the equilibrium constant $K^{\ddagger}$ for the formation of the activated complex (Eq. 44).

$$k_{rate}[A][B] = \frac{k_B T}{h} [AB^{\ddagger}] \tag{43}$$

$$K^{\ddagger} = \frac{[AB^{\ddagger}]}{[A][B]} = \frac{k_{rate} h}{T k_B} \tag{44}$$

Substituting the value of this equilibrium constant into Eq. 38,* and then the resulting expression for $\Delta G$ into Eq. 39, we find Eq. 45. Isolation of $\ln k_{rate}/T$ gives Eq. 46.**

$$-RT \ln \frac{k_{rate} h}{T k_B} = \Delta H^{\ddagger} - T\Delta S^{\ddagger} \tag{45}$$

$$\ln \frac{k_{rate}}{T} = -\frac{\Delta H^{\ddagger}}{RT} + \frac{\Delta S^{\ddagger}}{R} + \ln \frac{k_B}{h} \tag{46}$$

Since $\Delta H^{\ddagger}$ and $\Delta S^{\ddagger}$ are found to be nearly constant over a range of temperature, the only variables are $k_{rate}$ and $T$. Furthermore there is a linear relationship between $\ln k_{rate}/T$ and $1/T$ where the slope is $-\Delta H^{\ddagger}/R$ and the intercept is $\Delta S^{\ddagger}/R + \ln k_B/h$. Plotting these from a set of rate constants at several temperatures should give a straight line (Eyring plot), from which $\Delta H^{\ddagger}$ and $\Delta S^{\ddagger}$ may be obtained.† The line also indicates the quality of the data and allows extrapolations to predict rates at new temperatures.

Very similar results are obtained by simply plotting $\ln k_{rate}$ versus $1/T$ (Arrhenius plot). In this case the slope is $-E_a/R$ where $E_a$ is called the *Arrhenius activation energy*. For a reaction in solution, $E_a$ may be converted to $\Delta H^{\ddagger}$ by subtracting $RT$ (about 0.6 kcal at room temperature).[13]

For unimolecular reactions $\Delta S^{\ddagger}$ indicates whether the transition state has either more or less freedom of motion than the starting molecule. For example, the large positive $\Delta S^{\ddagger}$ for the thermal decomposition of

*The $K$ in Eq. 38 is numerically equal to the equilibrium constant, but it differs from the equilibrium constant in being dimensionless, owing to cancellation of concentration units in the complete form of Eq. 38 that results from integrating between limits:

$$\Delta G^e - \Delta G^0 = RT \ln \frac{[AB^{\ddagger}]^e [A]^0 [B]^0}{[AB^{\ddagger}]^0 [A]^e [B]^e}$$

This form is reduced to Eq. 38 by the fact that $\Delta G^e$ is zero at equilibrium, and standard state concentrations are all 1 $M$.

** $\dfrac{k_B}{h} = \dfrac{1.3803 \times 10^{-16} \text{ erg deg}^{-1} \text{ molecule}^{-1}}{6.6238 \times 10^{-27} \text{ erg sec molecule}^{-1}} = 2.0838 \times 10^{10} \text{ deg}^{-1} \text{ sec}^{-1}$

†We are accustomed to comparing $\Delta H^{\ddagger}$ and $\Delta S^{\ddagger}$ values of various reactions, but it is worth pointing out that the values are the amounts for converting a mole of each starting material to a mole of transition-state complex when the concentrations of starting materials and transition-state complex are all 1 molar; that is, $\Delta H^{\ddagger}$ and $\Delta S^{\ddagger}$ represent $\Delta H^{0\ddagger}$ and $\Delta S^{0\ddagger}$.

the azo compound in Eq. 47[14] is in accord with a simple fragmentation mechanism where the transition state is a more loosely bound, more freely moving structure than the starting molecule.

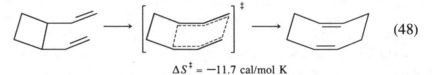

$$\Delta S^{\ddagger} = 16.3 \text{ cal/mol K}$$

In contrast with this, the large negative $\Delta S^{\ddagger}$ for the reaction in Eq. 48[15] requires that some freedom of motion in the starting molecule, such as rotation of the ring substituent bonds, is restricted in the transition state.

$$\Delta S^{\ddagger} = -11.7 \text{ cal/mol K}$$

The $\Delta S^{\ddagger}$ values calculated from second-order rate constants are not independent of the concentration units,[16] although they are generally negative. They can be used to compare similar reactions.

### 7.3.7  Substituent Effects on Kinetics

The order of bond breaking or bond forming in most reactions leads to temporary development of a positive or negative electrostatic charge at a particular site in the transition-state structure. Determining whether the charge is positive or negative and whether it is relatively large or small allows postulation of the sequence of bond changes. How can we determine this?

Certain substituents are known to withdraw electron density from a reaction site, thus rendering a developing negative charge at that site less intense and lessening the energy required to develop that negative charge. If a negative charge is partially formed on progressing from starting material to the transition state of a reaction, the substance with the electron-withdrawing substituent will react at a faster rate than one with a hydrogen atom. Other substituents are known to donate electron density to a reaction site, and these would intensify a developing negative charge, thus slowing the reaction. In other reactions a positive charge is developed at the reaction site. Here the substituents would have the opposite effect; that is, the electron-donating substituents will give faster reactions. For those reactions where the substituents are on a benzene ring and far enough from the reaction site to be sterically out of the way, the substituent effects have been correlated quantitatively. The Hammett equation (Eq. 49) states that the $\log_{10}$ of the ratio of the rate constant for some reaction with a substituent present ($k$) to the rate constant with no substituent ($k_0$) is equal to the product of a factor indicating the sensitivity of that reaction to substituent effects ($\rho$) and the characteristic electronic influence of the substituent ($\sigma$).[17]

$$\log_{10} \frac{k}{k_0} = \rho\sigma \qquad (49)$$

Table I lists the $\sigma$ values for a selection of common substituents in meta and para positions. The $\rho$ value for a reaction is determined by

**TABLE I. Selected Hammett $\sigma$ Values**

| Substitutent | $\sigma_{meta}$ | $\sigma_{para}$ | $\sigma^+$ |
|---|---|---|---|
| $NMe_2$ | −0.21 | −0.83 | −1.7 |
| $O^-$ | −0.17 | −0.52 | — |
| $NH_2$ | −0.16 | −0.66 | −1.3 |
| $CO_2^-$ | −0.10 | 0.00 | −0.03 |
| $CH_3$ | −0.07 | −0.17 | −0.31 |
| $C_2H_5$ | −0.07 | −0.15 | −0.30 |
| H | 0 | 0 | 0 |
| Ph | 0.06 | −0.01 | −0.17 |
| OH | 0.12 | −0.37 | −0.92 |
| $OCH_3$ | 0.12 | −0.27 | −0.78 |
| $SCH_3$ | 0.15 | 0.00 | −0.60 |
| $NHCOCH_3$ | 0.21 | 0.00 | 0.00 |
| OPh | 0.25 | −0.32 | −0.5 |
| SH | 0.25 | 0.15 | — |
| $CONH_2$ | 0.28 | — | — |
| F | 0.34 | 0.06 | −0.07 |
| I | 0.35 | 0.18 | 0.13 |
| CHO | 0.35 | 0.22 | — |
| Cl | 0.37 | 0.23 | 0.11 |
| COOH | 0.37 | 0.45 | 0.42 |
| $CO_2R$ | 0.37 | 0.45 | 0.48 |
| $COCH_3$ | 0.38 | 0.50 | — |
| Br | 0.39 | 0.23 | 0.15 |
| $OCOCH_3$ | 0.39 | 0.31 | — |
| $CF_3$ | 0.43 | 0.54 | — |
| CN | 0.56 | 0.66 | 0.66 |
| $NH_3^+$ | 0.63 | — | — |
| $NO_2$ | 0.71 | 0.78 | 0.79 |
| $NMe_3^+$ | 0.88 | 0.82 | 0.41 |

*Source:* Ritchie, C. D.; Sager, W. F. *Progr. Phys. Org. Chem.* **1964**, *2*, 323.

first measuring rate constants for a series of examples of the reaction with various substituents present. These are used to plot $\log_{10} k/k_0$ versus $\sigma$. The slope of this plot is $\rho$. With the value of $\rho$ and the table of $\sigma$ values, one may then predict $k$ values of other examples with different substituents.

Where did the $\sigma$ values come from? Hammett chose to define them from a particular reaction, the ionization of benzoic acid. In this case, and many others, the equilibrium constants rather than rate constants were measured. For equilibria the Hammett equation is written in terms of equilibrium constants:

$$\log_{10} \frac{K}{K_0} = \rho\sigma \tag{50}$$

Since the ionization of a series of substituted benzoic acids are used to define $\sigma$ values, the $\rho$ for this reaction was defined as 1.000. To find the $\sigma_m$ for the chloro substituent, the $K_a$ values for $m$-chlorobenzoic acid, $1.48 \times 10^{-4}$, and benzoic acid, $6.27 \times 10^{-5}$ are entered in Eq. 50 and it is solved for $\sigma$, which equals 0.373, the table value.

In the ionization of benzoic acid, a negative charge develops on the substituent-carrying molecule. Any other reaction in which a negative charge develops would be affected in the same direction by the substituents, and it would be found to have a positive value for $\rho$. In kinetic data, negative charge development in the transition state is indicated by the positive $\rho$. In reactions where a positive charge develops in the transition state or the product, the $\rho$ values will be negative. In reactions where the sensitivity to substituents is low, the $\rho$ can approach zero, and where it is high, the $\rho$ value can be as high as $\pm 5$ or more. This sensitivity is related to the distance separating the charge and the substituent, the transmitting ability of the intervening atoms, and also to the size of the charge, usually a fraction of a unit charge in transition states.

Estimates of the size of the charge in a transition state have been made by taking a ratio of $\rho$ values for a rate and an equilibrium in very similar reactions.[18]

In some reactions the electron-donating substituents in the para position can have a greater effect than their $\sigma$ values would predict. These are cases where a direct resonance contribution can be made as in benzylic carbocation formation:

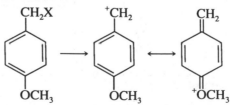

$$\tag{51}$$

For such reactions another set of substituent values, $\sigma^+$ was developed.[19] The rates of solvolysis of meta- and para-substituted cumyl chlorides in 90% aqueous acetone (Eq. 52) were measured. The $\rho$ for the reaction

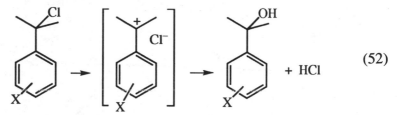

$$\text{(52)}$$

was found to be $-4.54$ by plotting only the meta isomers against $\sigma$. This was then used to obtain $\sigma^+$ values according to Eq. 53 with the para isomers. These are found in Table I, also. Notice that the $\sigma^+$ values are similar to the $\sigma_{para}$ values for the electron-withdrawing substituents, while the $\sigma^+$ are larger than the $\sigma_{para}$ for the electron-donating substit-uents, as would be expected for resonance contributions to the stability of carbocations.

$$\log_{10} \frac{k}{k_0} = -4.54\sigma^+ \qquad (53)$$

An investigation of the thermal decomposition of oxetanones is pre-sented here to illustrate the use of the Hammett equation. 2-Oxetanones decompose with first-order kinetics to give carbon dioxide and an alkene (Eq. 54).[20] A series of substituted 3- and 4-aryl-2-oxetanones was pre-pared and the rate constants determined for the decomposition at 150°C.

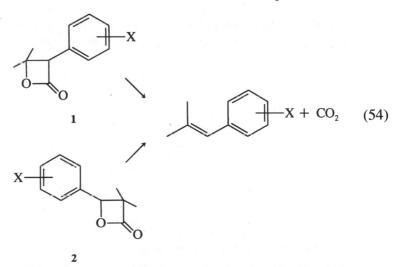

$$\text{(54)}$$

A plot of $\log k/k_0$ for each of these versus $\sigma^+$ is shown in Fig. X. The line for compound series **1** is almost flat ($\rho = +0.03$), indicating that

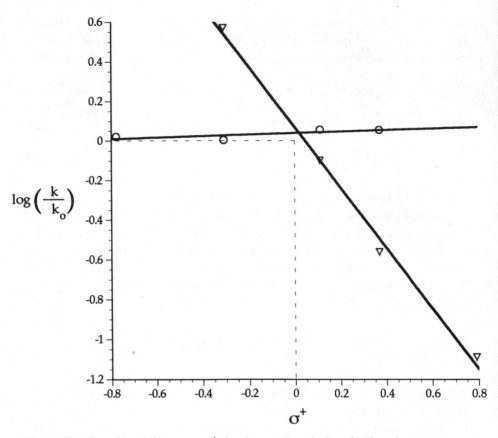

**Figure X.** Plot of log $k/k_0$ versus $\sigma^+$ for the reactions in Eq. 54 (3-aryloxetanones o; 4-aryloxetanones $\nabla$).

no significant charge develops at ring atom 3 in the transition state. The line for compound series **2** shows a substantial negative slope ($\rho = -1.52$), indicating that a partial positive charge develops at ring atom 4 in the transition state as indicated in **3**. The partial plus charge is at a

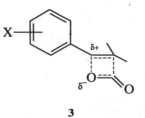

**3**

benzylic site, which is the reason for using $\sigma^+$ instead of $\sigma$. Thermolysis of *cis*-3-*tert*-butyl-4-phenyl-2-oxetanone gives pure *cis*-alkene, which suggests a concerted mechanism with some dipolar character, because a completely formed carbocation would be detached from the oxygen and likely to allow rotation of the $\sigma$ bond to give a mixture of *cis*-alkene and *trans*-alkene.

It is further possible that without a 4-aryl group, no significant charge would develop on ring atom 4 and the reaction would be more cleanly concerted, as allowed by the fact that compounds **1** all react faster than compounds **2**.

One may gain a qualitative view of the results without measuring $\rho$ or using a graph by noting that a partial positive charge develops at ring atom 4 as indicated by the faster rate for *p*-tolyl-**2** compared to phenyl-**2**. The electron-donating methyl group on the aromatic ring stabilizes the $\delta^+$ and facilitates its formation.

The familiar bromination of substituted benzenes with $Br_2$ in acetic acid also correlates with $\sigma^+$, and the $\rho$ is $-12$, indicating substantial plus-charge development very close to the substituents[21] as shown for anisole in Eq. 55. Notice also that the activating, ortho-, para-directing

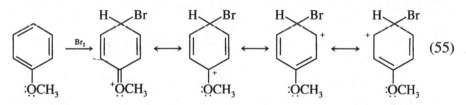

$$(55)$$

groups generally have negative $\sigma^+$ values and the deactivating, meta-directing groups have positive $\sigma^+$ values.

The $\rho$ values for an enormous number of reactions have been determined, and a lengthy list of $\sigma$ values is available. Beyond this, many variations on the original equation have been formulated.[22]

The Hammett equation is called a *linear free-energy relationship*. The log $K$ and log $k$ values are proportional to $\Delta G$ and $\Delta G^{\ddagger}$ values, respectively (Eq. 38) for equilibria and rates. The change in $\Delta G$ or $\Delta G^{\ddagger}$ with changing substituents for many reactions is linearly related to the $\sigma$ scale. Therefore, the change in $\Delta G$ or $\Delta G^{\ddagger}$ with changes in substituents in one reaction is linearly related to the change in $\Delta G$ or $\Delta G^{\ddagger}$ in another reaction for the same changes in substituents

$$\frac{\Delta \Delta G_1}{\rho_1} = \frac{\Delta \Delta G_2}{\rho_2} \tag{56}$$

where $\Delta \Delta G_1$ is the change in free energy for reaction 1 for a change in substituents, and $\Delta \Delta G_2$ is the change in free energy for reaction 2 for the same change in substituents.

## 7.4  REPRESENTATIVE MECHANISMS

After decades of mechanistic investigations and the discerning of generalities, it is now common practice to propose reasonable mechanisms for new reactions, considering their relationships to ones that have been investigated mechanistically. At the simplest level, these proposals consist of a succession of reaction intermediates without great attention to transition states. A sampling of such proposals is gathered in this section without supporting mechanistic data. An accounting of atoms and electrons must be included. For simplicity, only one resonance form of each ion or radical is given in most cases.

Some working principles to keep in mind in writing such proposals include the following:

1. Sites of unlike charge attract one another and often become bonded together.
2. Elements of different electronegativity that are bonded together carry partial charges.
3. In most ionic and concerted reactions, electrons remain paired throughout the process.
4. An odd, unpaired electron exists on radical species temporarily.
5. Complete electron octets are maintained on all C, N, O, or F atoms except at carbocation, radical, carbene, and nitrene sites where fewer than eight electrons reside temporarily.
6. In most steps, reactivity should decrease on progressing to product. For example highly basic compounds give products of lower basicity, and unstable 1° carbocations rearrange to 2° to 3° where possible.
7. Overall stability should increase. For example, conjugation may increase, especially aromatization, single bonds may form at the expense of the double bonds (as in the Diels–Alder reaction), and double bonds to oxygen may form at the expense of double bonds to carbon (as in the Claisen rearrangement).

### 7.4.1  Reactions in Basic Solution

In basic solutions, it is common for the base to remove a proton from carbon to give a nucleophilic species that attacks an electrophile to give new bonding. If the initial base is a carbanion generated by a redox process, it, too, attacks electrophiles. In other circumstances, a leaving group departs from the carbanion to generate an electrophile such as a carbene or benzyne, which then leads to other products. There are end-less variations of this, as may be appreciated by viewing many examples. A few are given here.

### A. *Overall reaction:*[23]

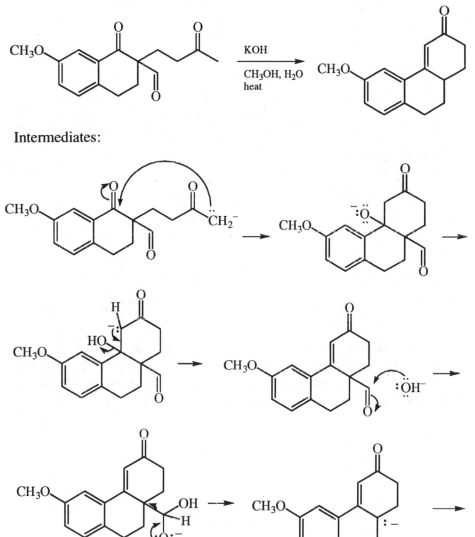

Intermediates:

Treatment of a ketone with the relatively weak base OH⁻ generates small equilibrium amounts of enolate anions at sites alpha to the carbonyl group. The nonbonding electron pair on the carbanion site is attracted to the partial plus on another ketone site and becomes a bond. The resulting alcohol eliminates readily even in basic solution in this case because conjugation is connected from the aromatic ring to the carbonyl group. The hydroxide attacks the formyl group directly with loss of formic acid to give an enolate ion that is delocalized to the α carbon and carbonyl oxygen. The final weak base present is formate.

### B. Overall reaction:[24]

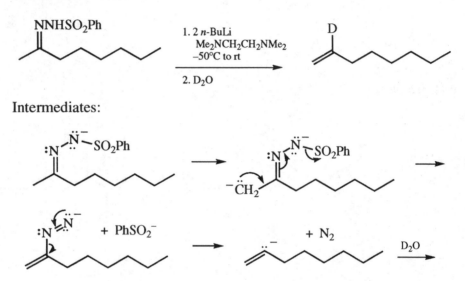

Intermediates:

Two equivalents of strong base remove a proton from nitrogen and another from an α carbon. The weak base phenylsulfinate departs, followed by the very stable molecule $N_2$. The resulting vinyl carbanion is then deuterated.

### C. Overall reaction:[25]

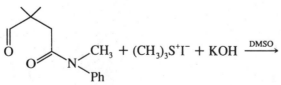

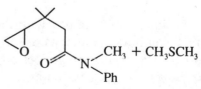

Intermediates:

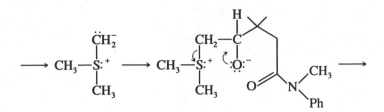

The hydroxide removes a proton from the trimethylsulfonium ion to give a 1,2-dipolar species called an *ylide*, which has nucleophilic character on the $CH_2$ group. This adds to an aldehyde to give an alkoxide that displaces the dimethyl sulfide leaving group.

### D. Overall reaction:[26]

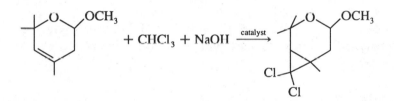

Intermediates:

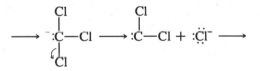

The hydroxide removes a proton from chloroform to give the trichlorocarbanion which loses a chloride ion to give the neutral dichlorocarbene. This is electrophilic and forms two σ bonds to the alkene site.

Some other electrophiles that convert alkenes to cyclopropanes, are not free carbenes but have metals coordinated with their electrophilic site. These are called *carbenoids*, and include the Simmons–Smith reaction and the copper-, rhodium-, or palladium-catalyzed decomposition of diazoketones and esters (Section 5.3). That the metal atom is present in the electrophile is shown by the variation of the stereoselectivity of the reaction with changes in the other ligands on metal.

### E. Overall reaction:[27]

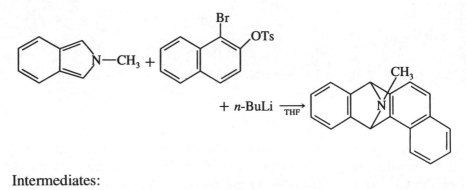

Intermediates:

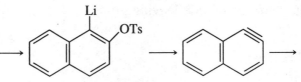

The *n*-butyllithium exchanges with the bromo compound to give *n*-butyl bromide and an aryllithium. Elimination of lithium tosylate gives the naphthalyne, which combines with the isoindole in a Diels–Alder reaction (Section 8.3.2).

### F. Overall reaction:[28]

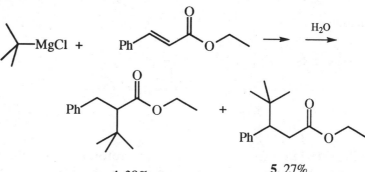

**4**, 38%          **5**, 27%

Intermediates leading to **4**:

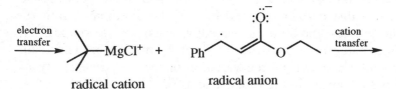

radical cation          radical anion

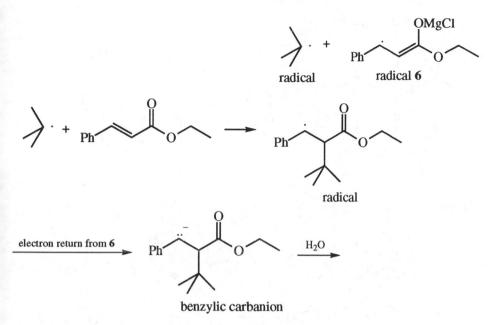

radical

benzylic carbanion

Product **5**, but not **4**, would be expected from a carbanion mechanism via the more stable enolate intermediate. However, product **4** might arise from the more stable benzylic free radical. This is a single-electron-transfer mechanism where the easily reduced cinnamate ester gains a single electron from the Grignard reagent to become a radical anion, leaving the Grignard reagent as a radical cation. Transfer of the magnesium cation leaves a *tert*-butyl radical that attacks another ethyl cinnamate molecule alpha to the ester to give a benzylic radical. This radical then takes back the electron originally transferred to the first cinnamic acid, giving the benzylic carbanion, which gives product **4** on protonation. Product **5** may arise from competing conventional carbanion conjugate addition or from the combination of the cinnamate radical anion with the *tert*-butylmagnesium radical cation. Addition of α-methylstyrene to the original reaction mixture suppresses the formation of **4** by capturing the *tert*-butyl radicals, indicating that it is a radical process.

## 7.4.2  Reactions in Acidic Solution

In acid solutions, generally a proton or other Lewis acid attaches to an electron-rich site, giving a carbocation that may lead on to a series of other carbocations, often ending finally with loss of the Lewis acid or a proton. The carbocations are frequently stabilized by resonance with much of the charge residing on an oxygen atom.

## A. *Overall reaction:*[29]

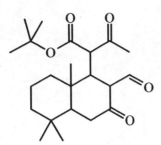  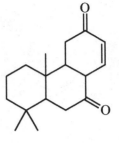

Intermediates:

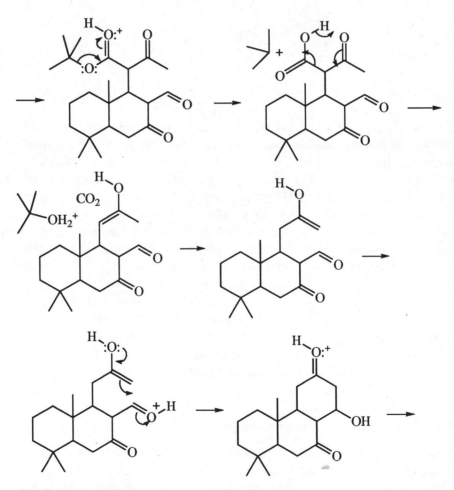

Some minor variations that may parallel this proposal include showing the β-ketoaldehyde as the enol form and then protonating it on carbon instead of oxygen. Both may be involved and it is of little consequence.

**B. Overall reaction:**[30]

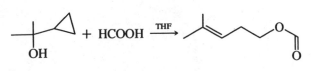

Intermediates:

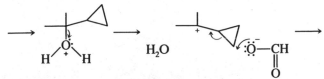

More or fewer carbocation intermediates could be drawn here. The last step is drawn as a concerted loss of one bond and gain of two others. The amount of concerted versus discrete steps is still a matter of debate in many cases.

**C. Overall reaction:**[31]

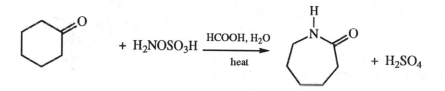

Intermediates:

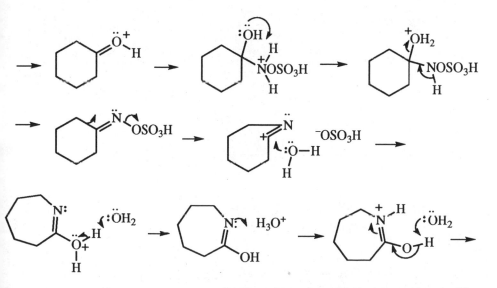

This is a variation on the classic Beckmann rearrangement of oximes.[32]

### D. Overall reaction:[33]

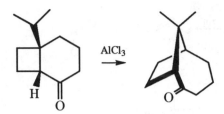

Intermediates:

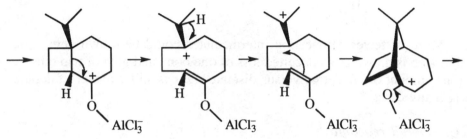

## 7.4.3  Free-Radical Reactions

Certain classes of compounds will decompose thermally or photochemically in a homolytic way; that is, a bonding pair of electrons will unpair.
Aliphatic azo compounds and peroxides commonly do so and can be
used in catalytic amounts to initiate radical chain reactions of other
molecules.

### A. Overall reaction:[34]

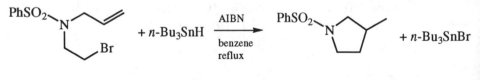

AIBN is $(CH_3)_2CCNN=NCCN(CH_3)_2$

Intermediates:

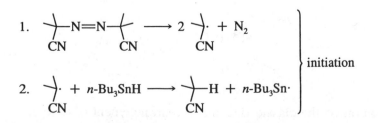

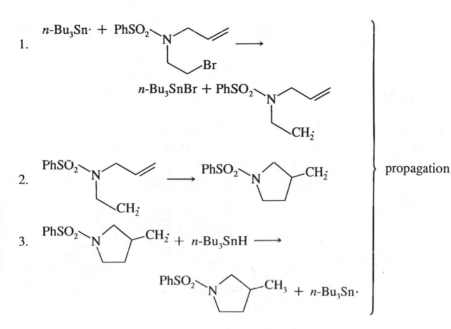

1. propagation
2.
3.

### 7.4.4 Rearrangement to Electron-Deficient Nitrogen

The Lossen and Hoffmann rearrangements involve an α elimination from an amide nitrogen to leave the nitrogen neutral but short of an octet of electrons. The loss of $N_2$ in the Curtius and Schmidt rearrangements gives a similar circumstance on nitrogen.

**A. Overall reaction:**[35]

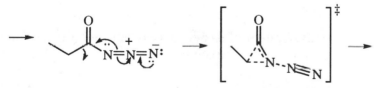

Intermediates:

The alkyl group on the other side of the carbonyl group migrates with bonding electrons to make up the deficiency developing on nitrogen. In other cases where the migrating carbon was a chiral center, configuration was retained. This may well be a concerted process where the electron deficiency on nitrogen is no more than slight and the bracketed structure is only a transition state with no intermediates between the acyl azide and the isocyanate.

## PROBLEMS

1. Using linear graphical analysis of data, the slope of the best line can be evaluated by selecting two convenient, far-apart points on the line (not data points) and calculating $\Delta H^{\ddagger}$ or $\rho$ using equations such as the following. Derive these equations from Eqs. 46 and 49 in this chapter.

$$\ln \frac{k_1}{k_2} + \ln \frac{T_2}{T_1} = \frac{\Delta H^{\ddagger}}{R} \left( \frac{1}{T_2} - \frac{1}{T_1} \right)$$

$$\log \frac{k_1}{k_2} = \rho(\sigma_1 - \sigma_2)$$

2. The rates of reaction of substituted benzoyl chlorides with excess methanol at 0°C were measured. The pseudo-first-order rate constant for *meta*-chlorobenzoyl chloride is 0.0877 moles/L sec, and for *para*-nitrobenzoyl chloride is 0.416 moles/L sec. Calculate the rate constant for *meta*-methylbenzoyl chloride, assuming these few examples give a good correlation.[36] (Of course, in a proper study a larger set of examples would be used to obtain a correlation.)

3. Imidoyl chlorides hydrolyze to give amides as you might expect from their structural similarity to acid chlorides.[37] The pseudo-first-

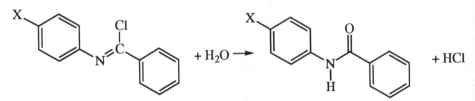

order rate constants for hydrolysis in excess aqueous dioxane at 25°C for three examples are tabulated:

$$X = -NO_2 \qquad 3.5 \times 10^{-4} \ sec^{-1}$$

$$X = -Cl \qquad 112. \times 10^{-4} \ sec^{-1}$$

$$X = -H \qquad 460. \times 10^{-4} \ sec^{-1}$$

a. Without taking the time to plot a graph, assume that these are typical substituents and they give a good Hammett correlation; calculate the Hammett $\rho$ for the reaction.

b. Calculate a predicted rate constant for the hydrolysis of

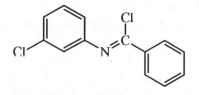

c. Added chloride ion in the solution showed a pronounced retarding effect on the rate of hydrolysis. Propose a mechanism for the reaction in accord with this and the $\rho$ value.

4. The rate of the following reaction at three concentrations of NaOH in aqueous acetonitrile at 0°C was determined from absorbance measurements at 294 nm. The initial concentration of **A** was $1.13 \times 10^{-4}$ M. The results for three runs are tabulated, where time is in seconds.[38]

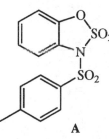

$$+ 2\,NaOH \longrightarrow$$

$$+ H_2O$$

| 0.0500 M NaOH | | 0.0600 M NaOH | | 0.1000 M NaOH | |
|---|---|---|---|---|---|
| Time | Absorbance | Time | Absorbance | Time | Absorbance |
| 0 | 0.145 | 0 | 0.155 | 0 | 0.162 |
| 10 | 0.160 | 10 | 0.170 | 4 | 0.175 |
| 20 | 0.178 | 20 | 0.183 | 10 | 0.190 |
| 30 | 0.190 | 30 | 0.195 | 18 | 0.205 |
| 40 | 0.200 | 40 | 0.205 | 26 | 0.217 |
| 60 | 0.219 | 60 | 0.218 | 40 | 0.230 |
| 80 | 0.230 | 80 | 0.225 | ∞ | 0.253 |
| ∞ | 0.259 | ∞ | 0.241 | | |

Evaluate the data graphically, and calculate pseudo-first-order rate constants for each run. What is the overall order of the reaction? Write a rate expression for the reaction, and calculate the actual rate constant for the reaction. This is a very fast reaction. How many seconds elapsed between the combining of the reagents and time 0 of the tabulated data?

5. Diphenylazomethane ($PhCH_2N=NCH_2Ph$) reacts at elevated temperature to give nitrogen gas, among other products.[39] The kinetics of this reaction were studied as follows. A closed constant volume flask containing diphenyl ether solvent was thermally equilibrated in a constant temperature bath. A sample of the azo compound was then injected into the container and the gradual pressure increase measured. The cumulative increase in pressure ($\Delta P$) at a list of times is tabulated below for two reaction temperatures.

| 150.0°C | | 175.0°C | |
|---|---|---|---|
| Time (min) | $\Delta P$ (mm Hg) | Time (min) | $\Delta P$ (mm Hg) |
| 0.0 | 0.0 | 0.0 | 0.0 |
| 8.0 | 5.5 | 2.0 | 18.3 |
| 16.0 | 10.2 | 4.0 | 31.7 |
| 24.0 | 14.5 | 6.0 | 40.7 |
| 32.0 | 18.0 | 8.0 | 47.2 |
| 40.0 | 21.1 | 10.0 | 51.6 |
| 48.0 | 24.0 | 12.0 | 54.6 |
| 56.0 | 26.2 | $\infty$ | 61.8 |
| 64.0 | 28.3 | | |
| 72.0 | 30.0 | | |
| $\infty$ | 40.4 | | |

a. How can you determine whether this is a first-order or second-order reaction from these data?

b. Find the value of the rate constant $k$ graphically at each temperature. *Hint:* The total amount of A ($PhCH_2N=NCH_2Ph$) or the initial concentration of A is proportional to the total amount of $N_2$ obtained and to $\Delta P_\infty$. The amount of A remaining at time $t$ is proportional to $\Delta P_\infty - \Delta P_t$. The ratio of $[A]_0/[A]_t$ then equals $\Delta P_\infty/(\Delta P_\infty - \Delta P_t)$.

c. Calculate the activation enthalpy and entropy for the reaction.

**6.** The following reaction occurs in $CCl_4$ at 25°C:        Ref. 40

$$CH_3-\underset{\underset{CH_3}{|}}{\overset{\overset{I}{|}}{C}}-\underset{\underset{CH_3}{|}}{\overset{\overset{Cl}{|}}{C}}-CH_3 + ICl \longrightarrow CH_3-\underset{\underset{CH_3}{|}}{\overset{\overset{Cl}{|}}{C}}-\underset{\underset{CH_3}{|}}{\overset{\overset{Cl}{|}}{C}}-CH_3 + I_2$$

Kinetic measurements were made following the concentration of ICl, using a large excess of $C_6H_{12}ICl$. The data fit a linear plot as shown below. Write a rate expression for the reaction and calculate the rate constant.

|  | Initial Concentrations (M) | |
| --- | --- | --- |
|  | Run 1 | Run 2 |
| $[C_6H_{12}ICl]^0$ | 0.0349 | 0.0174 |
| $[ICl]^0$ | 0.00441 | 0.00160 |

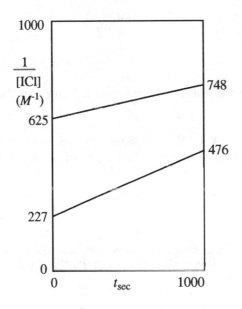

**7.** Using the kinetic data tabulated below, determine the order and write the rate expression for the illustrated reaction, rounding to the nearest whole order(s). Explain how you did this. Calculate the rate constant including the units.[41]

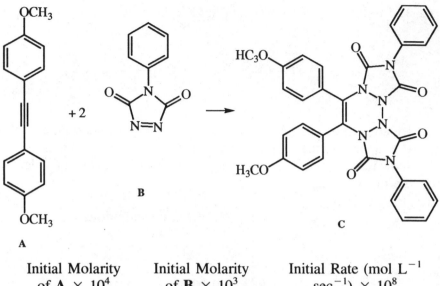

| Initial Molarity of A $\times 10^4$ | Initial Molarity of B $\times 10^3$ | Initial Rate (mol L$^{-1}$ sec$^{-1}$) $\times 10^8$ |
|:---:|:---:|:---:|
| 3.85 | 9.95 | 7.78 |
| 1.96 | 10.3 | 4.18 |
| 0.99 | 10.2 | 2.04 |
| 1.96 | 4.87 | 1.89 |
| 1.96 | 6.74 | 2.73 |

Which of the following reaction schemes fit the preceding results? Explain.

| | |
|---|---|
| **A + B → I** slow | **B + B → I** slow |
| **I + B → C** fast | **I + A → C** fast |
| | |
| **A + B → I** fast | **A + B + B → C** |
| **I + B → C** slow | |

8. The rate of the following reaction was measured with varying concentrations of $PdCl_4^{2-}$, $H^+$, and $Cl^-$, and the results are given in the table.[42]

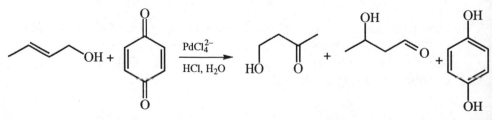

| Run[a] | $[PdCl_4^{2-}]$ | $[H^+]^b$ | $[Cl^-]^c$ | $10^5 k_{obs}$, sec$^{-1}$ |
|---|---|---|---|---|
| 1 | 0.005 | 0.2 | 0.6 | 0.52 |
| 2 | 0.0125 | 0.2 | 0.6 | 1.7 |
| 3 | 0.025 | 0.2 | 0.6 | 2.7 |
| 4 | 0.050 | 0.2 | 0.6 | 5.1 |
| 5 | 0.025 | 0.2 | 0.4 | 5.1 |
| 6 | 0.025 | 0.2 | 0.9 | 1.5 |
| 7 | 0.025 | 0.2 | 1.2 | 0.77 |
| 8 | 0.025 | 0.4 | 0.6 | 1.6 |
| 9 | 0.025 | 0.6 | 0.6 | 1.2 |
| 10 | 0.025 | 0.8 | 0.6 | 0.84 |

[a] All runs are in aqueous solution at 25°C. LiClO$_4$ was added to bring the ionic strength to 2.0. Initial 2-buten-1-ol and quinone concentrations are 0.005 molar.
[b] Added as HClO$_4$.
[c] Added as LiCl.

The tabulated $k_{obs}$ is from the pseudo first-order expression

$$- \frac{d[A]}{dt} = k_{obs}[A]$$

where A is 2-buten-1-ol. The benzoquinone simply oxidizes the Pd(0) back to Pd(II) to maintain a constant concentration of Pd(II). Note the effect of changing concentrations on the observed rate, and write a rate expression for the oxidation of A in terms of all involved concentrations (except benzoquinone, which doesn't affect the rate). Round off all orders to the nearest whole number. Select a run from the table and calculate an actual rate constant for the reaction giving units for it. (*Hint*: see Eq. 19.)

Draw likely intermediates in the mechanism of each of the reactions (in problems 9–22).

**9.**

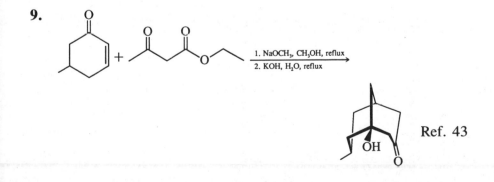

1. NaOCH$_3$, CH$_3$OH, reflux
2. KOH, H$_2$O, reflux

Ref. 43

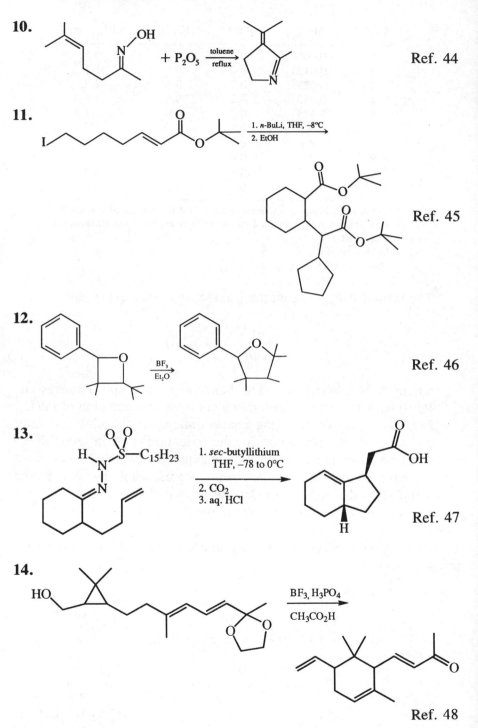

**10.**                                        Ref. 44

**11.**                                        Ref. 45

**12.**                                        Ref. 46

**13.**                                        Ref. 47

**14.**                                        Ref. 48

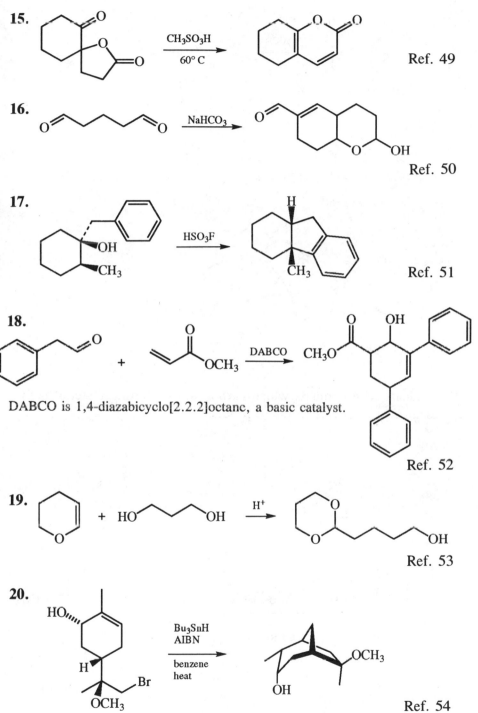

**15.**

$$\xrightarrow[60°\,C]{CH_3SO_3H}$$

Ref. 49

**16.**

$$\xrightarrow{NaHCO_3}$$

Ref. 50

**17.**

$$\xrightarrow{HSO_3F}$$

Ref. 51

**18.**

+

$$\xrightarrow{DABCO}$$

DABCO is 1,4-diazabicyclo[2.2.2]octane, a basic catalyst.

Ref. 52

**19.**

+   HO⌒OH

$$\xrightarrow{H^+}$$

Ref. 53

**20.**

$$\xrightarrow[\substack{benzene\\heat}]{\substack{Bu_3SnH\\AIBN}}$$

Ref. 54

**21.**

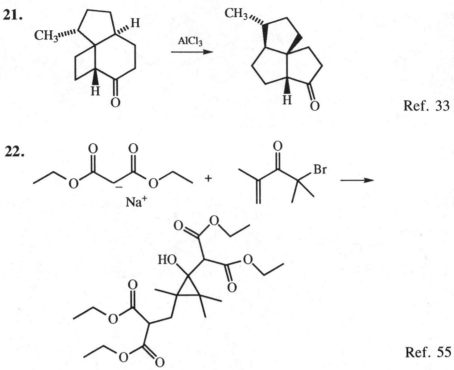

Ref. 33

**22.**

Ref. 55

**23.** Indicate with numbers which carbon in the starting material is the source of each carbon in the product:

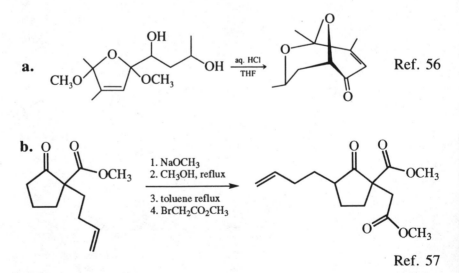

**a.**    aq. HCl / THF    Ref. 56

**b.**    1. NaOCH₃  2. CH₃OH, reflux  3. toluene reflux  4. BrCH₂CO₂CH₃    Ref. 57

24. Treatment of 1-chloro-2-methylcyclohexene with 5 eq of methyl-lithium in TMEDA–THF followed by aqueous workup gave a 47% yield of 1-methylbicyclo[4.1.0]heptane.[58] Two mechanisms were proposed for this process as outlined below. What experiment(s) would you use to determine which mechanism is incorrect? Tell what you would do, what results you might expect, and how you would draw you conclusions.

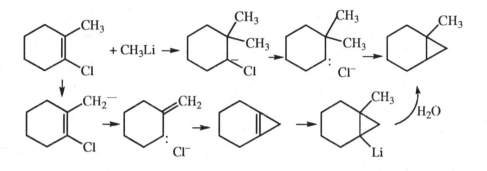

25. The reaction of monoolefins with nitroso compounds gives hydroxylamine products.[59] The process was thought to occur by either a one-step ene reaction (Eq. 1, below) or a two-step reaction via a transient aziridine oxide intermediate (Eq. 2). The following deuterium isotope effects were measured for the reaction of 2,3-dimethyl-2-butene and hexafluoronitrosobenzene. Explain the similarities and differences among these results in terms of one of these mechanisms. Use the results to exclude one mechanism. The first

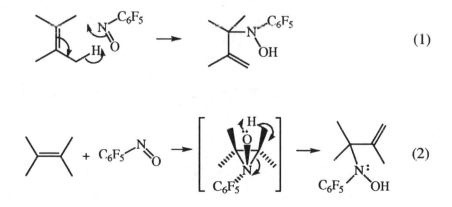

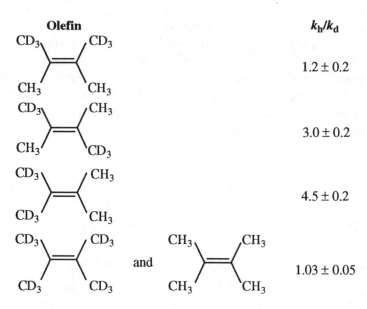

three cases are intramolecular competitions where $k_H$ is the rate of formation of the $C=CH_2$ product and $k_D$ is the rate of formation of the $C=CD_2$ alternative product. The last case is an intermolecular competition. The results were determined by $^1H$ NMR analysis of the products.

26. Vinyl ethers react rapidly with 2,2-bis(trifluoromethyl)ethylene-1,1-dicarbonitrile to afford cyclobutenes.[60] Mechanisms may be proposed, including a one-step concerted process where both $\sigma$ bonds form simultaneously, or two-step processes via one of the following dipolar intermediates:

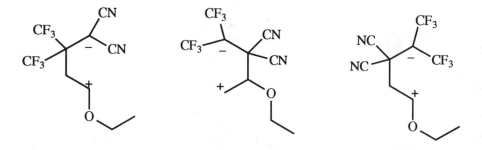

Considering the following facts, eliminate all except one mechanism. Explain.

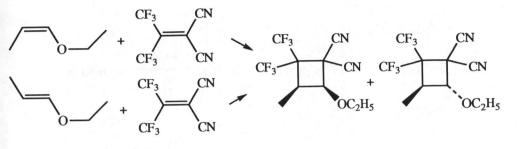

| In benzene the cis propenyl ether gives | 87 | : | 13 |
| In benzene the trans propenyl ether gives | 10 | : | 90 |
| In acetone the cis propenyl ether gives | 82 | : | 18 |
| In acetone the trans propenyl ether gives | 39 | : | 61 |

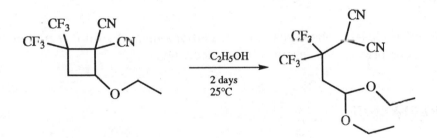

Relative rates of reaction with 2,2-bis(trifluoromethyl)ethylene-1,1-dicarbonitrile

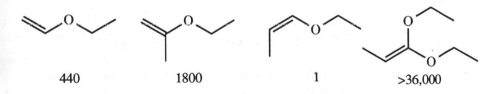

| 440 | 1800 | 1 | >36,000 |

**27.** Explain in terms of mechanism, why one of the following sulfonates is more reactive:[61]

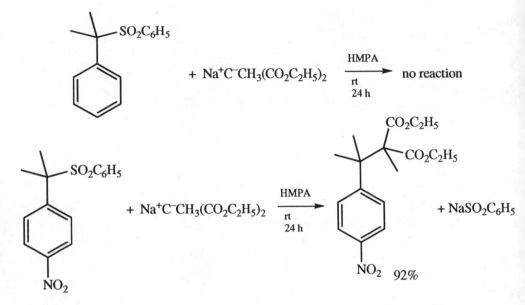

In other closely related reactions, small amounts of sulfur or *p*-dinitro-benzene prevent the reaction from occurring.

# REFERENCES

1. Carpenter, B. K. *Determination of Organic Reaction Mechanisms*, Wiley-Interscience, New York, 1984.
2. Sykes, P. *The Search for Organic Reaction Pathways*, Wiley, New York, 1972.
3. Hammond, G. S. *J. Am. Chem. Soc.* **1953**, *77*, 334.
4. Mendelson, W.; Pridgen, L.; Holmes, M.; Shilcrat, S. *J. Org. Chem.* **1989**, *54*, 2490.
5. Johri, K. K.; DesMarteau, D. D. *J. Org. Chem.* **1983**, *48*, 242.
6. Bridges, A. J.; Whitham, G. H. *J. Chem. Soc. Chem. Commun.* **1974**, 142.
7. Moore, J. W.; Pearson, R. G. *Kinetics and Mechanism*, 3rd ed., Wiley-Interscience, New York, 1981.
8. Wigfield, D. C.; Gowland, F. W. *J. Org. Chem.* **1980**, *45*, 653.
9. Nome, F.; Erbs, W.; Correia, V. R. *J. Org. Chem.* **1981**, *46*, 3802.
10. Wilkins, R. G. *The Study of Kinetics and Mechanism of Reactions of Transition Metal Complexes*, Allyn & Bacon, Boston, 1974, p. 27.
11. Melander, L.; Saunders, Jr., W. H. *Reaction Rates of Isotopic Molecules*, Wiley-Interscience, New York, 1980.

12. Rastetter, W. H.; Wagner, W. R.; Findeis, M. A. *J. Org. Chem.* **1982,** *47,* 419.

13. Moore, J. W.; Pearson, R. G. *Kinetics and Mechanism,* 3rd ed., Wiley-Interscience, New York, 1981, pp. 178–181.

14. Martin, J. C.; Timberlake, J. W. *J. Am. Chem. Soc.* **1970,** *92,* 978.

15. Hammond, G. S.; DeBoer, C. *J. Am. Chem. Soc.* **1964,** *86,* 899.

16. Robinson, P. J. *J. Chem. Ed.* **1978,** *55,* 509.

17. Hammett, L. P. *J. Am. Chem. Soc.* **1937,** *59,* 96.

18. Poh, B.-L. *Can. J. Chem.* **1979,** *57,* 255.

19. Brown, H. C.; Okamoto, Y. *J. Am. Chem. Soc.* **1958,** *80,* 4979.

20. Imai, T.; Nishida, S. *J. Org. Chem.* **1979,** *44,* 3574.

21. Brown, H. C.; Okamoto, Y. *J. Am. Chem. Soc.* **1958,** *80,* 4979.

22. Ritchie, C. D.; Sager, W. F. *Progr. Phys. Org. Chem.* **1964,** *2,* 323.

23. Turner, R. B.; Nettleton, Jr., D. E.; Ferebee, R. *J. Am. Chem. Soc.* **1956,** *78,* 5923.

24. Stemke, J. E.; Bond, F. T. *Tetrahedron Lett.* **1975,** 1815.

25. Majewski, M.; Snieckus, V. *J. Org. Chem.* **1984,** *49,* 2684.

26. Slessor, K.; Oehlschlager, A. C.; Johnston, B. D.; Pierce, Jr., H. D.; Grewal, S. K.; Wickremesinghe, L. K. G. *J. Org. Chem.* **1980,** *45,* 2290.

27. Gribble, G. W.; LeHoullier, C. S.; Sibi, M. P.; Allen, R. W. *J. Org. Chem.* **1985,** *50,* 1611.

28. Holm, T.; Crossland, I.; Madsen, J. Ø. *Acta Chem. Scand. B* **1978,** *32,* 754.

29. Meyer, W. L.; Manning, R. A.; Schroeder, P. G.; Shew, D. C. *J. Org. Chem.* **1977,** *42,* 2754.

30. Moiseenkov, A. M.; Czeskis, B. A.; Nefedov, O. M. *Synthesis* **1985,** 932.

31. Olah, G. A.; Fung, A. P. *Synthesis* **1979,** 537.

32. Gawley, R. A. *Org. React.* **1988,** *35,* 1.

33. Kakiuchi, K.; Fukunaga, K.; Matsuo, F.; Ohnishi, Y.; Tobe, Y. *J. Org. Chem.* **1991,** *56,* 6742.

34. Padwa, A.; Nimmesgern, H.; Wong, G. S. K. *J. Org. Chem.* **1985,** *50,* 5620.

35. Kricheldorf, H. R.; Leppert, E. *Synthesis* **1976,** 329.

36. Norris, J. F.; Young, H. H. Jr. *J. Am. Chem. Soc.* **1935,** *57,* 1420.

37. Hegarty, A. F.; Cronin, J. D.; Scott, F. L. *J. Chem. Soc. Perkin Trans. II,* **1975,** 429.

38. Andersen, K. K.; Bray, D. D.; Chumpradit, S.; Clark, M. E.; Habgood, G. J.; Hubbard, C. D.; Young, K. M. *J. Org. Chem.* **1991,** *56,* 6508.

39. Bandlish, B. K.; Garner, A. W.; Hodges, M. L.; Timberlake, J. W. *J. Am. Chem. Soc.* **1975,** *97,* 5856.

40. Schmid, G. H.; Gordon, J. W. *J. Org. Chem.* **1983,** *48,* 4010.

41. Cheng, C.-C.; Greene, F. D.; Blount, J. F. *J. Org. Chem.* **1984,** *49,* 2917.

42. Zaw, K.; Henry, P. M. *J. Org. Chem.* **1990,** *55,* 1842.

43. Kraus, G. A.; Hon, Y.-S. *J. Org. Chem.* **1985,** *50,* 4605.

44. Gawley, R. E.; Termine, E. J. *J. Org. Chem.* **1984,** *49,* 1946.

45. Cooke, M. P., Jr. *J. Org. Chem.* **1984,** *49,* 1144.

46. Carless, H. A. J.; Trivedi, H. S. *J. Chem. Soc. Chem. Commun.* **1979,** 382.

47. Chamberlin, A. R.; Bloom, S. H.; Cervini, L. A.; Fortsch, C. H. *J. Am. Chem. Soc.* **1988,** *110,* 4788.

48. Chopra, A. K.; Khambay, B. P. S.; Madden, H.; Moss, G. P.; Weedon, B. C. L. *J. Chem. Soc. Chem. Commun.* **1977,** 357.

49. Mandal, A. K.; Jawalkar, D. G. *J. Org. Chem.* **1989,** *54,* 2364.

50. Tashima, T.; Imai, M.; Kuroda, Y.; Yagi, S.; Nakagawa, T. *J. Org. Chem.* **1991,** *56,* 694.

51. Barrow, C. J.; Bright, S. T.; Coxon, J. M.; Steel, P. J. *J. Org. Chem.* **1989,** *54,* 2542.

52. Poly, W.; Schomburg, D.; Hoffmann, H. M. R. *J. Org. Chem.* **1988,** *53,* 3701.

53. Kozikowski, A. P.; Stein, P. D. *J. Org. Chem.* **1984,** *49,* 2301.

54. Srikrishna, A.; Hemamslini, P. *J. Org. Chem.* **1990,** *55,* 4883.

55. Barbee, T. R.; Hedeel, G.; Heeg, M. J.; Albrizati, K. F. *J. Org. Chem.* **1991,** *56,* 6773.

56. DeShong, P.; Ramesh, S.; Perez, J. J.; Bodish, C. *Tetrahedron Lett.* **1982,** *23,* 2243.

57. Dauben, W. G.; Walker, D. M. *Tetrahedron Lett.* **1982,** *23,* 711.

58. Gassman, P. G.; Valcho, J. J.; Proehl, G. S.; Cooper, C. F. *J. Am. Chem. Soc.* **1980,** *102,* 6519.

59. Seymour, C. A.; Greene, F. D. *J. Org. Chem.* **1982,** *47,* 5226.

60. Huisgen, R.; Brückner, R. *Tetrahedron Lett.* **1990,** *31,* 2553, 2557.

61. Kornblum, N.; Ackermann, P.; Manthey, J. W.; Musser, M. T.; Pinnick, H. W.; Singaram, S.; Wade, P. A. *J. Org. Chem.* **1988,** *53,* 1475.

# 8

# ELECTRON DELOCALIZATION, AROMATIC CHARACTER, AND PERICYCLIC REACTIONS

Alternating behavior has been observed in many different kinds of reactions. Protonation of a conjugated polyene gives a carbocation with the positive charge specifically delocalized to alternating carbons of the chain. For example, protonation of hexatriene gives positive charge on carbons 1, 3, and 5 as manifest in final product distributions and in spectra. Similar distributions exist in conjugated carbanions.

Another alternation appears in the series of cyclic molecules that show aromatic character. If the ring contains a conjugated cycle of $4n + 2$ $\pi$ electrons, we find peculiar spectroscopic and reaction characteristics that are absent in the $4n$ set ($n$ is the series 0, 1, 2, 3, . . .).

In this chapter you will see reactions that occur when the reactant(s) contain(s) $4n + 2$ $\pi$ electrons, and different reactions that occur when there are $4n$ $\pi$ electrons.

Recognition of these alternating patterns has allowed many predictions. In this chapter these patterns are rationalized in terms of molecular orbitals.

## 8.1 MOLECULAR ORBITALS

Matter has a dual character, showing the properties of particles and of waves. Diffraction of electromagnetic radiation gives patterns of interference and reinforcement caused by overlapping waves that are either out of phase or in phase, establishing the wave nature of radiation.

Similar patterns of diffraction have been produced by directing a beam of electrons through a crystal of nickel, showing the wave nature of electrons also. Electrons confined by their electrostatic charge to the very small volume of an atom or molecule are best described as standing waves, rather than as particles. Each standing wave is approximated mathematically by a wavefunction that gives the shape and energy of the wave. The higher the energy state of the electron, the more nodes in the wave. The square of the wavefunction at any location around the nucleus is directly proportional to the electron density at that location.

For a $p$ orbital the wavefunction values are high at two points on opposite sides of the nucleus and diminish with distance from these points, to zero at a plane through the nucleus or at infinite distance from the atom. The sign of the function (phase of the wave) is opposite on each side of that zero (nodal) plane through the nucleus. On paper we can show a cross section of this with contour lines as shown in Fig. I. The square of this is the electron density function. A plot of the amplitude of the $2p_y$ Slater orbital of fluorine along the $y$ axis is shown in Fig. II. The square of this is shown also to indicate the relative electron density along that axis. In the fluorine molecule the molecular orbitals are conveniently described as combinations of the two atomic $p$ orbitals. They overlap in two ways: with reinforcement in the space between the nuclei ($\sigma$ orbital) or with interference between the nuclei ($\sigma^*$ orbital). This is illustrated in Fig. III, where you can see that the electron density between nuclei in the $\sigma^*$ orbital would be very small and the nuclei would repel each other. However, the electron density is largely con-

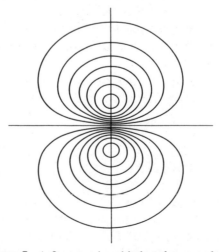

**Figure I.** A $2p_y$ atomic orbital at the $x$, $y$ plane.

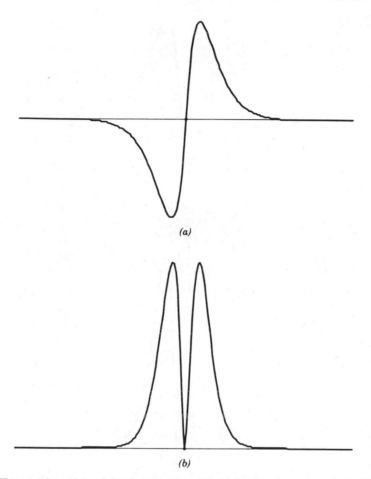

**Figure II.** (*a*) The value of the $2p_y$ atomic orbital of a fluorine atom ($\psi_{2p_y}$) along the $y$ axis, and (*b*) the square of that function indicating electron density along that axis. The nucleus is at the center and the full trace width is 10 Å.

centrated between the nuclei in the $\sigma$ orbital, and the nuclei are attracted to it and, therefore, together. These are the antibonding and bonding orbitals. Of course the electron density in the *volume* between the nuclei is important, but for simplicity we have sampled that along the axis in these plots.

The bonding electron pair in $F_2$ occupies the lower-energy $\sigma$ orbital, and the $\sigma^*$ is unoccupied in the ground state.

In ethylene, besides the $\sigma$ orbitals, there are $\pi$ molecular orbitals described as laterally overlapping $p$ orbitals. Here again there are two ways of overlapping: with reinforcement ($\pi$) and with interference ($\pi^*$) in the space above and below the nodal plane. A conventional cross-

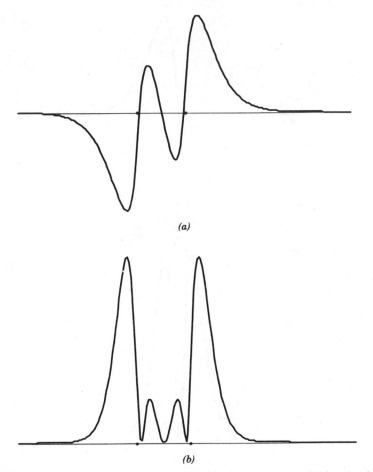

**Figure III.** Molecular orbital values and electron densities along the internuclear axis in the $F_2$ molecule, based on overlap of $\psi_{2p_y}$ at the normal bonding distance. (a) $\sigma^*$, (b) $\sigma^{*2}$, (c) $\sigma$, (d) $\sigma^2$. These plots were provided by S. L. Whittenburg, University of New Orleans.

sectional representation of these is shown in Fig. IV where opposite phases are indicated by solid and dotted contours. The bonding pair of electrons resides in the lower-energy $\pi$ orbital where the attractive forces effectively bond the carbon atoms together. The change in energy when one electron in an isolated carbon $p$ orbital spreads to occupy an ethylene $\pi$ orbital is called $\beta$ ($-18$ kcal/mol). The $\pi$-bonding energy $B_\pi$ of ethylene with two electrons is $2\beta$ according to the Hückel approximation where the interelectron repulsion is neglected. Conjugated systems consist of three or more $p$ orbitals, where the bonding energy is more than $\beta$ for each electron in the lowest bonding molecular orbital. As more $p$

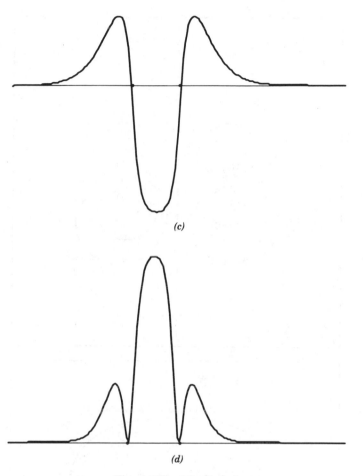

*(c)*

*(d)*

**Figure III.** (*Continued*)

orbitals are mixed in a linear array a matching number of $\pi$ molecular orbitals are formed and the lowest energy orbital is lower, approaching a minimum of $2\beta$ per electron (Figure V). Each molecular orbital is composed of portions of the atomic $p$ orbitals, and the extent to which each atomic orbital contributes to each molecular orbital is given as a coefficient. This is called a linear combination of atomic orbitals (LCAO).[1,2] The square of the coefficient is the electron density at that atom if one electron occupies the molecular orbital with that coefficient. In butadiene the sum of the squares of all the coefficients on carbon 1 (or any other carbon in Fig. V) is one atomic orbital. Likewise, the sum of the squares of the coefficients of the four atoms contributing to one molecular orbital is also one. In each set of molecular orbitals shown in

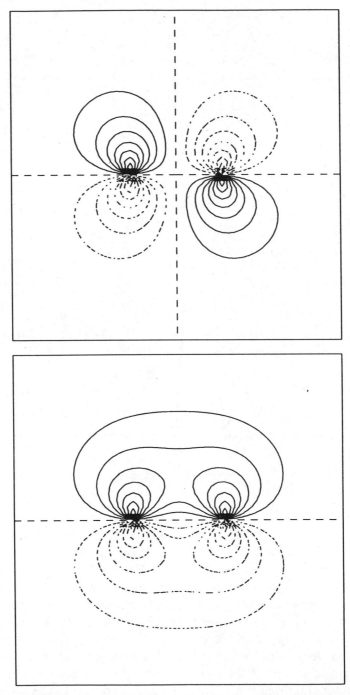

**Figure IV.** Bonding ($\pi$) and antibonding ($\pi^*$) molecular orbitals of ethylene. Opposite phases are indicated by solid and dotted contours. These plots were provided by E. A. Boudreaux, University of New Orleans.

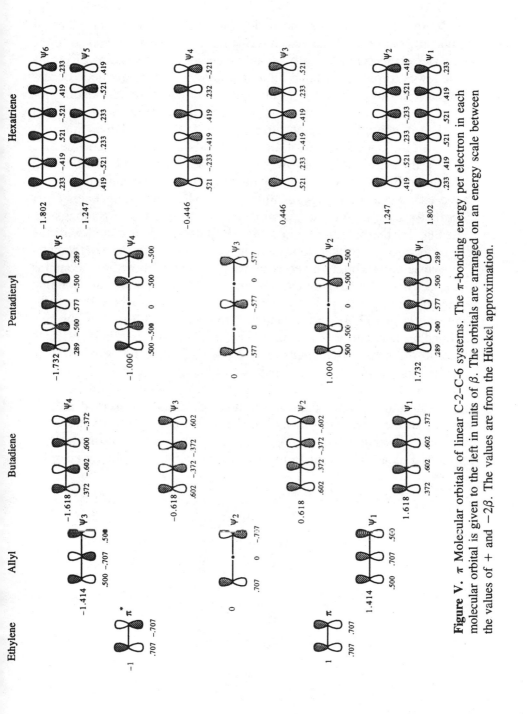

**Figure V.** $\pi$ Molecular orbitals of linear C-2–C-6 systems. The $\pi$-bonding energy per electron in each molecular orbital is given to the left in units of $\beta$. The orbitals are arranged on an energy scale between the values of $+$ and $-2\beta$. The values are from the Hückel approximation.

235

Fig. V, the lowest ($\psi_1$) has no nodes (beyond the one in the plane of the nuclei), the next-higher ($\psi_2$) has one node, $\psi_3$ has two, and so forth. The nodes occur symmetrically about the center of the system. Electrons in the lowest orbital will bond all atoms together. Where a node occurs between a pair of carbons, electron occupation will give antibonding for that pair. Where a node occurs at an atom, the coefficient is zero and there is a nonbonding relationship with the flanking carbons. For simplicity, all the contributions of atomic orbitals are drawn the same size and in a distorted small narrow shape in the figure. The net bonding molecular orbitals are shown below the zero-energy line and antibonding above on a scale of energy in units of $\beta$. The $\pi$-bonding energy per electron for each molecular orbital is also given in units of $\beta$ alongside.

In Fig. VI the butadiene molecular orbitals are represented with the size of the atomic orbital contributions proportional to the coefficients. The shapes approximate a series of waves. Neutral butadiene contains four electrons, which populate $\psi_1$ and $\psi_2$. Those in $\psi_1$ are bonding throughout and particularly so between C-2 and C-3, where the coefficients are large. Those in $\psi_2$ bond C-1 to C-2 and C-3 to C-4 but give antibonding between C-2 and C-3. This antibonding is relatively weak owing to the small coefficients at C-2 and C-3. The net $\pi$ bonding in butadiene is greater than in a pair of isolated ethylenes, which is in accord with the observation of greater thermodynamic stability in conjugated systems compared to nonconjugated. This is observed experimentally in the lower heat of hydrogenation of conjugated dienes compared to nonconjugated analogs and in the selective formation of conjugated dienes in elimination reactions. Using the $\pi$-bonding energies

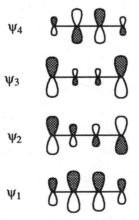

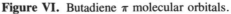

**Figure VI.** Butadiene $\pi$ molecular orbitals.

per electron given in Fig. V, we find that two ethylenes with two electrons each have less bonding energy ($4\beta$) than does butadiene with four electrons in $\psi_1$ and $\psi_2$ ($2 \times 1.618\beta + 2 \times 0.618\beta = 4.472\beta$). The additional $0.472\beta$, the *delocalization energy*, is the value of conjugation. The orbitals above the zero $\beta$ line are net antibonding, and if they were occupied, the molecule would weaken.

The orbital bonding energies also explain the higher kinetic reactivity of conjugated systems over nonconjugated. The electrons in the highest occupied molecular orbital (HOMO) are the most easily moved to new bonding relationships and can be considered the molecular valence electrons.[3] The antibonding part of $\psi_2$ of butadiene places it $0.382\beta$ higher energy (closer to the zero-energy line) than the HOMO of ethylene.

The systems containing an odd number of conjugated carbons are reactive intermediate radicals, carbocations, and carbanions. The allyl radical contains three $\pi$ electrons evenly spread to all three carbons (no charges). The allyl carbocation has two electrons in $\psi_1$ where the coefficient is high on the central carbon; therefore, the plus charge (electron deficiency) will be on the first and third carbons. The charge at each carbon is calculable by taking the squares of the coefficients at that carbon and multiplying each square by the number of electrons in that molecular orbital. Summing these for the orbitals that are occupied gives the total $\pi$ electron density on that atom. If it is 1.00, the atom is neutral; if less than 1.00, the shortage is the plus charge; and if more than 1.00, the excess is the negative charge. This is shown for all carbons in the pentadienyl cation and anion in Fig. VII. The result is that the plus charge in the cation is evenly divided by 3, with $\frac{1}{3}$ on each of the alternating carbons 1, 3, and 5. The anion is divided in the same way. This is the first alternation rationalized by molecular orbitals.

Conjugation is more effective in stabilizing species with odd numbers of overlapping $p$ orbitals, as reflected in delocalization energies. The total $\pi$-bonding energy in the allyl carbanion is $2 \times 1.414\beta + 2 \times 0\beta = 2.818\beta$. Comparing this with the energy of four electrons in the double bond and isolated $p$ orbital ($2\beta$) leaves $0.818\beta$ as the delocalization energy. This is greater than that of butadiene. This can rationalize the much greater reactivity of allyl chloride compared to propyl chloride toward metals such as magnesium, where negative charge develops on carbon.

In Section 8.3 reactions of polyenes are examined with particular attention to the phase relationships in the highest occupied (HOMO) and lowest unoccupied molecular orbitals (LUMO). Anticipating this, notice that for the neutral polyenes the HOMO is a row of alternating bonding

| | C-1 | C-2 | C-3 | C-4 | C-5 | |
|---|---|---|---|---|---|---|
| $\psi_3$ | 0.577 | 0 | −0.577 | 0 | 0.577 | Coefficients |
| $\psi_3^2$ | 0.333 | 0 | 0.333 | 0 | 0.333 | |
| $2\psi_3^2$ | 0.666 | 0 | 0.666 | 0 | 0.666 | |
| $\psi_2$ | 0.500 | 0.500 | 0 | −0.500 | −0.500 | Coefficients |
| $\psi_2^2$ | 0.250 | 0.250 | 0 | 0.250 | 0.250 | |
| $2\psi_2^2$ | 0.500 | 0.500 | 0 | 0.500 | 0.500 | |
| $\psi_1$ | 0.289 | 0.500 | 0.577 | 0.500 | 0.289 | Coefficients |
| $\psi_1^2$ | 0.0835 | 0.250 | 0.333 | 0.250 | 0.0835 | |
| $2\psi_1^2$ | 0.167 | 0.500 | 0.667 | 0.500 | 0.167 | |
| $2\psi_1^2 + 2\psi_2^2$ | 0.667 | 1.000 | 0.667 | 1.000 | 0.667 | Electron density |
| $1.00 - 2\psi_1^2 + 2\psi_2^2$ | +0.333 | 0 | +0.333 | 0 | +0.333 | Charges on cation |
| $2\psi_1^2 + 2\psi_2^2 + 2\psi_3^2$ | 1.333 | 1.000 | 1.333 | 1.000 | 1.333 | Electron density |
| $1.000 - (2\psi_1^2 + 2\psi_2^2 + 2\psi_3^2)$ | −0.333 | 0 | −0.333 | 0 | −0.333 | Charges on anion |

**Figure VII.** Calculation of the charge on each atom of the pentadienyl cation and anion.

and antibonding relationships whatever the length (shaded 2 up, 2 down, 2 up, etc.), and that the $C_{4n}$ have the opposite phase at the first and last carbons while the $C_{4n+2}$ have the same phase at the first and last carbons.

## 8.2 AROMATIC CHARACTER

The lowest molecular orbital in the linear polyenes approaches a $\pi$-bonding energy of $2\beta$ as the chain becomes longer and the end carbons become a smaller fraction of the conjugated system. If the system is cyclic and conjugated all the way around, there are no ends and the lowest becomes $2\beta$ for all ring sizes. As with acyclic cases there are as many molecular orbitals as there were contributing atomic orbitals. Unlike acyclic compounds however, there are degenerate (equal-energy) pairs of molecular orbitals above the lowest level. These orbitals are diagrammed on a $\pi$-bonding energy scale in Fig. VIII for rings with up to eight members.

Each vertical step involves an additional nodal plane, which cuts the ring at two sites. In the ground state the electrons would populate the lower energy orbitals following Hund's rule. In cyclobutadiene the lowest energy pair of electrons give bonding among all the carbons. However the next two electrons are in nonbonding orbitals, contributing nothing to the stability of the molecule. In the tricyclic dimer of cyclobutadiene, all electrons are in bonding orbitals; therefore, cyclobutadiene is an unstable transient molecule that rapidly dimerizes. In benzene or the isoelectronic furan, all the $\pi$ electrons are in orbitals of net bonding, giving a stable system. In cyclooctatetraene (if it were conjugated) two electrons would occupy nonbonding orbitals. This again would be unstable and it is relieved by not conjugating. The molecule is tub-shaped, and the $\pi$ electrons are all in bonding "ethylene" orbitals, avoiding the angle strain present in a planar octagonal ring. These circumstances alternate through the cyclic vinylogous series continuing to larger rings. Those with $4n + 2$ $\pi$ electrons have the highest occupied energy level filled with four electrons (closed-shell occupation) and, if these are all bonding orbitals, they have greater stability than found in the acyclic cases. The greater stability in the $4n + 2$ series is called *aromatic character* and is manifest in numerous physical and chemical properties. They give a lower combustion energy compared to acyclic analogs and resist addition reactions that would interrupt the ring of $\pi$ bonding. The $^1H$ NMR spectra show differences, also. Hydrogens projecting outward in the plane of an aromatic ring appear several parts per million to the left of the vinyl hydrogens in acyclic analogs. Contrari-

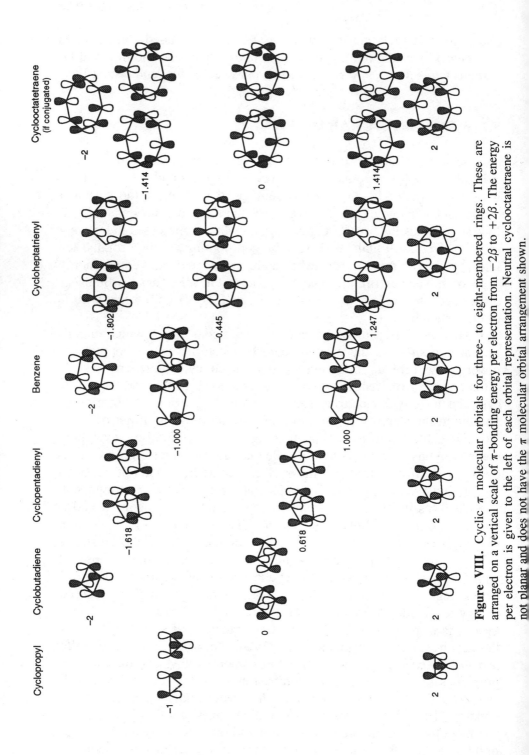

**Figure VIII.** Cyclic $\pi$ molecular orbitals for three- to eight-membered rings. These are arranged on a vertical scale of $\pi$-bonding energy per electron from $-2\beta$ to $+2\beta$. The energy per electron is given to the left of each orbital representation. Neutral cyclooctatetraene is not planar and does not have the $\pi$ molecular orbital arrangement shown.

wise, those hydrogens extending inward or above the ring appear considerably to the right.

Among the odd-membered rings, the aromatic ions are readily prepared. Cyclopentadiene is deprotonated by alkoxide bases while cycloheptatriene is not, even with stronger bases. On the other hand, bromocycloheptatriene is ionic while 5-bromocyclopentadiene is not. Tripropylcyclopropenyl perchlorate exists largely as the carbocation in aqueous acetonitrile at pH $\leq 7$.[4]

Molecules with $4n$ $\pi$ electrons in a ring of $p$ orbitals have a partially filled energy level, and the four and five membered rings have less stability than the acyclic analogs. They are called *antiaromatic*.[5] The stability differences can be seen by calculating and comparing the $\pi$ bonding energy for cyclic and acyclic compounds with the same number of $\pi$ electrons using the orbital energy values in Figs. V and VIII. The $(CH)_{16}$ and $(CH)_{24}$ rings have been prepared and show the opposite $^1H$ NMR chemical-shift effects from those found for the $4n + 2$ compounds.[6]

At the lowest energy level in carbocyclic compounds there is a single orbital, and above this the degenerate pairs begin. Thus in $4n + 2$, the 2 fills the lowest level and the $4n$ fill succeeding levels. Other numbers give incomplete levels. Those monocyclic aromatic systems where $n$ is greater than 1 are not nearly as chemically resistant as benzene, although some of them undergo electrophilic substitution, and some show large NMR shifts. For example, the outer protons on $(CH)_{18}$ appear at 8.8 ppm and the inner at $-1.8$ ppm. Fused-ring polycyclic compounds with $4n + 2$ $\pi$ electrons show aromatic character, but there is less stabilization per ring than in benzene.[7]

Some heterocyclic compounds show aromatic character despite their lack of degenerate orbitals. This may involve one $\pi$ electron from the heteroatom as in pyridine, or two as in pyrrole and furan, making a set of 6 $\pi$ electrons in a ring.[8] Many fused-ring heterocyclic compounds are aromatic also. A 10 $\pi$ electron monocyclic example, 1,4-dioxocin does not show aromatic character.[9]

There is some value in closed-shell electron occupation even if four of the electrons are nonbonding. Cyclooctatetraene can be reduced to the dianion by alkali metals. X-Ray crystallography of 1,3,5,7-tetramethylcyclooctatetraene dianion shows that the ring carbons form an equilateral planar octagon.[10] If it were not planar, the two extra electrons would reside in antibonding orbitals of alkene subunits, rather than the nonbonding $\psi_4$ and $\psi_5$ of flat cyclooctatetraene.

This is the second alternation, rationalized as having closed shell occupation of degenerate bonding molecular orbitals.

## 8.3  PERICYCLIC REACTIONS

Numerous organic reactions proceed through a cyclic transition state in a single step involving no intermediates. These are called *pericyclic reactions*.[11] Certain bonds break, while others form in concert. These reactions are divided into two groups taking stereochemical choices into account. Examples from one group have been observed under thermal conditions, while examples from the other group are found only under photochemical conditions. In the thermally observable reactions the phase relationships of the combining orbitals are such that the bonding electrons maintain bonding character from starting materials, through the transition state, and in the products. These concerted processes are therefore favored over nonconcerted ones, where bonds must first be broken to give radicals or other intermediates, which subsequently give new bonds. A smaller activation energy is necessary when the exothermal formation of the new bonds is under way during the endothermal breaking of the original bonds. These concerted reactions are stereospecific, while reactions involving preliminary breakage of bonds often lose stereochemical integrity.

Selection rules were put forth by Woodward and Hoffmann based on the conservation of orbital symmetry.[12] Molecular orbitals in the starting material are correlated with orbitals in the product according to their symmetry elements. If all the ground-state occupied molecular orbitals in the starting material correlate with ground-state occupied molecular orbitals in the product, the reactions will be allowed thermally. If, instead, a ground-state occupied orbital in the starting material correlates with an antibonding orbital in the product, photochemical excitation will be necessary. In photochemically allowed reactions, the lowest excited state of the starting material correlates with the lowest excited state of the product. It is the HOMO that is pivotal in these choices.

Fukui examined the interaction of the HOMO and LUMO alone (the frontier orbitals) and rationalized the same rules.[13, 14] Basically each reaction is viewed as the coalescing (or dissociation) of two sets of molecular orbitals intra- or intermolecularly. The HOMO of one is matched with the LUMO of the other, and if the overlap at both sites of projected new bond formation between them is in-phase (a bonding overlap), the reaction is allowed. This method of analysis is detailed with examples in Sections 8.3.1–8.3.3.

Another approach, concentrating on the transition state is given in Section 8.3.4.

All these analyses view the reactions as reversible and allowedness applies both ways. Other factors such as ring strain, steric hindrance, and bond energies determine the energetically favorable direction.

### 8.3.1  Electrocyclic Reactions

An electrocyclic reaction[15] is the closure of a conjugated polyene to give a cyclic compound with one less $\pi$ bond, or the reverse. For the first example, consider the opening of a cyclohexadiene to give a hexatriene (Eq. 1). A $\sigma$ bond breaks, and new $\pi$ bonding develops concurrently.

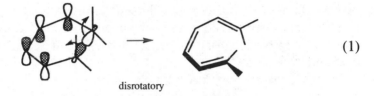

disrotatory

(1)

The highest energy electrons available are in the HOMO of the diene, the phase of which is indicated. This may donate electron density to the LUMO of the $\sigma$ bond, the phase of which is also shown. It can be seen that a particular rotating motion will bring the orbitals into the new bonding relationship with bonding overlap at both joining sites. Viewed from the side, one appears to rotate about 90° counterclockwise while the other goes clockwise. This is called *disrotatory motion*. The opposite, where both rotate clockwise or both counterclockwise is called *conrotatory motion*. These are illustrated in a general way in Fig. IX. The other arbitrary choice of phase indication in the initially independent

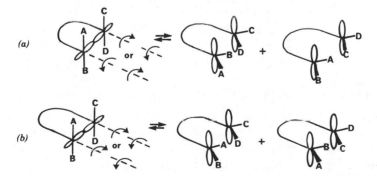

**Figure IX.** (*a*) Disrotatory and (*b*) conrotatory ring opening or closing.

molecular orbitals is as shown in Eq. 2. Both choices are equally valid

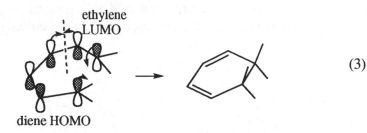

$$\text{disrotatory} \tag{2}$$

for consideration of allowedness and both indicate disrotatory motion, but each may lead to different stereoisomeric products. Two allowed motions should be considered in all pericyclic reactions. Which of the two disrotatory motions occurs will be determined in most cases by steric or electronic effects among the substituents, or both may occur. Conrotation is forbidden for this reaction because it leads to antibonding overlap at one end.

In this analysis we looked only at the closest-lying LUMO and HOMO, which are called the *frontier orbitals*. If we had instead examined the HOMO of the $\sigma$ bond and the LUMO of the diene, we would have again concluded that the disrotatory process was the allowed one.

The conclusion applies to the reverse of the reaction as well. The closure of the six-membered ring is very favorable, and there are more examples of that than of opening. We can analyze the closure by arbitrarily viewing the hexatriene as a diene and ethylene system about to interact (Eq. 3). The HOMO of the diene and the LUMO of the ethylene

ethylene
LUMO

$$\tag{3}$$

diene HOMO

can overlap in phase (bonding) if the motion is disrotatory (just as we concluded for the closure). The stereochemistry was demonstrated in the cyclization of the isomers of 2,4,6-octatriene.[16] The *E,Z,E* isomer gave only the cis ring compound. The *Z,Z,E* and *Z,Z,Z* isomers were in thermal equilibrium, but rate measurements indicate that the *Z,Z,E* isomer likely gave the trans ring compound (Eq. 4). The ring opening

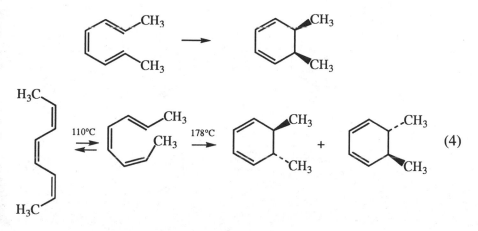

(4)

of cyclobutenes can be analyzed by the frontier orbital theory, also. The LUMO of the σ bond and the HOMO of the π bond are shown:

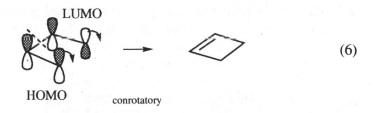

(5)

In contrast to the cyclohexadiene case, conrotatory motion is necessary here for in-phase overlap. As in the previous case, we can invert the phase indication of one of the molecular orbitals and see conrotatory motion in the other direction. We can also examine the closure viewing the diene as two ethylenes combining (LUMO of one with HOMO of the other):

LUMO

HOMO          conrotatory

(6)

Both directions of conrotation in the opening of *cis*-3,4-dimethylcy-clobutene lead to *Z,E*-2,4-hexadiene (Eq. 7).[17] The *trans*-3,4-dimethyl-cyclobutene might lead to the *E,E* and *Z,Z* dienes, but only the *E,E* was found. There is a strong tendency for electron-donating groups to rotate outward (Section 8.3.4).

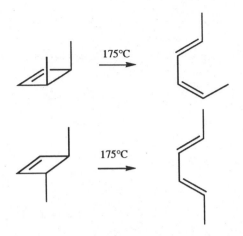

(7)

Another experiment showed both directions of conrotation for both opening and closure, and the complete absence of disrotation.[18] Each

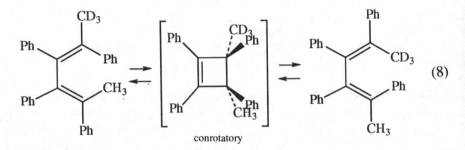

conrotatory

(8)

isomer of the deuterium labeled diene was heated at 124°C and a 1:1 mixture of them was obtained, containing none of the cis,cis isomer. In another experiment the trans,trans and the cis,cis dienes (no deuterium) were thermally equilibrated and none of the cis,trans diene was formed.

Higher vinylogs continue the alternation. The closure of a tetraene to a cyclocotatriene (or the reverse) can be analyzed as above, and one finds that conrotation is the thermally allowed process.

Under photochemical conditions we find the opposite results. Photons generally excite an electron from the HOMO to the next-higher molecular orbital. This higher orbital was the LUMO, but it becomes the highest singly occupied molecular orbital (HSOMO). In butadiene, for example, the ground-state HOMO was $\psi_2$ (Fig. V), but after photoexcitation, the HSOMO is $\psi_3$. This orbital can then interact with the LUMO of the other reacting system to give the product. The phase relationship of the HSOMO will be the opposite of the former HOMO since there is one more node; therefore, the stereochemistry will be opposite that of the ground-state thermal reactions. For example, irradiation of *cis-*

5,6-dimethyl-1,4-diphenylcyclohexadiene gave the conrotatory triene product (Eq. 9).[19] The butadiene $\psi_3$ interacts with the $\sigma$ LUMO to give the ring-opened triene, where in-phase overlap at both joining sites required a conrotatory motion. The low temperature was necessary to avoid a subsequent sigmatropic shift (see Section 8.3.3).

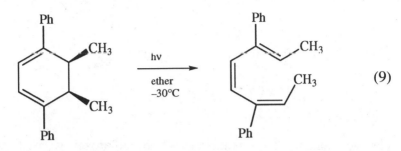

$$(9)$$

We can summarize the selection rules for allowed electrocyclic reactions as an alternating series:

| Closures | Openings | Thermal | Photochemical |
|---|---|---|---|
| Dienes | Cyclobutenes | Conrotatory | Disrotatory |
| Trienes | Cyclohexadienes | Disrotatory | Conrotatory |
| Tetraenes | Cyclooctatrienes | Conrotatory | Disrotatory |
| Pentaenes | Cyclodecatetraenes | Disrotatory | Conrotatory |
| Hexaenes | Cyclododecapentaenes | Conrotatory | Disrotatory |

Structural constraints can prevent reactions that would be allowed by these rules. For example, the very strained bicyclo[2.2.0]hex-2-ene opens slowly at 130°C to cyclohexadiene, while the less strained bicyclo[4.2.0]octa-2,4-diene is in rapid equilibrium with cyclooctatriene at 60°C. The allowed conrotatory movement in the first case would lead to cyclohexadiene with a trans double bond, which is impossible. However, the latter opening is disrotatory, giving all-cis double bonds in the cyclooctatriene. The higher-temperature reaction, disallowed by these rules, is a higher activation energy process, perhaps involving radicals, but not concerted.

## 8.3.2  Cycloaddition Reactions

A cycloaddition reaction is the joining together of two independent $\pi$-bonding systems to form a ring with two new $\sigma$ bonds. The reverse is called a *retrocycloaddition* reaction, and the selection rules again apply in both directions of a given reaction.

If butadiene and an appropriately substituted ethylene approach and begin to overlap as in Eq. 10, we find that there is a favorable phase relationship between the HOMO of one and the LUMO of the other for a face-to-face joining. This is, of course, the familiar Diels–Alder reaction, and it is thermally allowed:

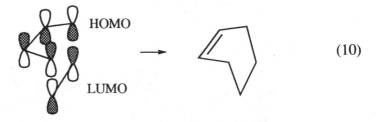

$$\text{(10)}$$

The stereochemical consequences of this approach were illustrated in Section 6.6.

The gap between the HOMO of butadiene and the LUMO of ethylene (or the LUMO of butadiene and the HOMO of ethylene) $1.618\beta$ (Fig. V) is too great for favorable interaction, and no reaction occurs. Substituents can narrow this gap. Electron-withdrawing substituents lower the energy level of LUMOs and electron-donating substituents raise the energy level of HOMOs. Thus the lower-energy LUMO in maleic anhydride is close enough in energy to the HOMO of butadiene for favorable reaction.[20] An electron-rich alkene such as ethyl vinyl ether has a higher-energy HOMO than ethylene, and will react with electron-poor dienes as in Eq. 37 of Chapter 5. Here the closer frontier orbitals are the LUMO of the diene and the HOMO of the dienophile. These are called *reverse-electron-demand* reactions. Either way, the phase relationships predict a thermally allowed reaction.

You should recall from descriptions of electrophilic aromatic substitution that stronger electron-donating substituents have nonbonding electron pairs on the attaching atom, while electron-withdrawing (acceptor) substituents have a positive charge on the attaching atom.

In unsymmetric cases, the regiochemistry may be predicted from the frontier orbital coefficients at the joining atoms.[21] An electron-donating substituent on carbon 1 of a diene or alkene will enlarge the coefficient of the HOMO at the remote end. An electron acceptor group on carbon 1 will enlarge the coefficient of the LUMO at the remote end (Fig. X). In the reaction, the greater orbital overlap interaction occurs when the atoms with larger coefficients overlap with each other and those with the smaller coefficients overlap with each other, as opposed to overlaps of smaller with larger. Thus in Diels–Alder reactions where a donor is

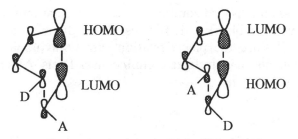

**Figure X.** Regioselectivity in Diels–Alder reactions.

on carbon 1 of the diene and an acceptor is on carbon 1 of the alkene, the donor and acceptor will selectively appear on adjacent cyclohexene ring atoms. The same is true when donor and acceptor are on alkene and diene, respectively. A donor on carbon 2 of a diene enlarges the HOMO coefficient at carbon 1, and an acceptor on carbon 2 enlarges the LUMO coefficient at carbon 1. This favors a 1,4-disubstituted cyclohexene product. The increase in regioselectivity brought about with Lewis acid catalysis is a result of coordination of the catalyst with a receptor substituent, making it a stronger receptor. These preferences were originally given in Section 6.6. The calculation of these coefficients is beyond the scope of this book.

If two ethylenes are brought together face to face, the HOMO–LUMO phase relationship is unfavorable for formation of a cyclobutane (Eq. 11). Under photochemical conditions, however, this is a valuable route

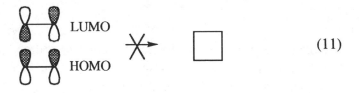

$$\tag{11}$$

to cyclobutanes (Section 6.4). The allowedness may be determined by considering the phase relationship of the HSOMO with the LUMO as in Eq. 12. Those substituents that were cis on an ethylene will be cis on the cyclobutane. Regioisomers are possible, also.

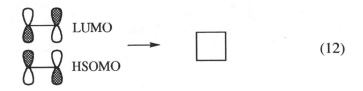

$$\tag{12}$$

A triene and an ethylene combining face to face at their ends is again not thermally favorable (Eq. 13), as is also a diene combining with a diene (Eq. 14) since the phase relations are unfavorable to bonding. These components may instead combine in a Diels–Alder reaction to give a vinylcyclohexene.

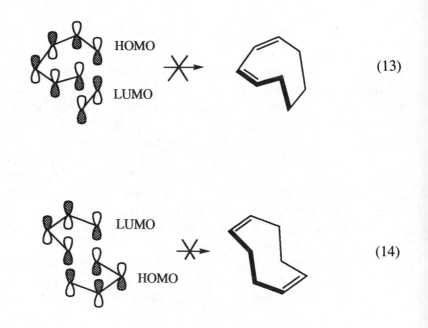

(13)

(14)

Going one step further, the combination of a diene and a triene (Eq. 15), or a tetraene and an ethylene, is favorable face to face. Specific examples are shown in Eq. 16[22] and Eq. 17.[23]

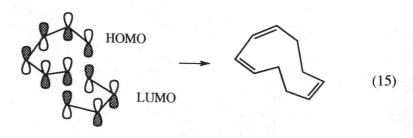

(15)

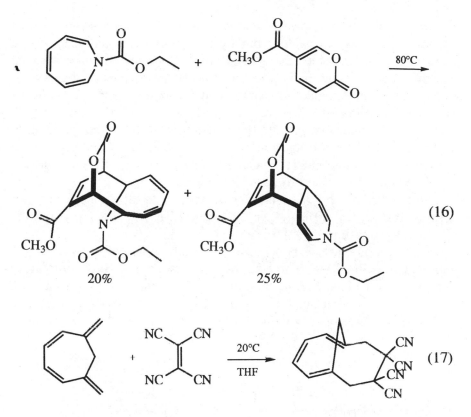

$$(16)$$

20%          25%

$$(17)$$

These face-to-face reactions are in effect syn additions for both $\pi$ systems in the same sense as syn hydroxylation or epoxidation of an alkene. Groups cis on the alkene remain cis on the ring. In contrast to these, the addition of bromine to alkenes is an anti addition. For example, cyclohexene gives the trans dibromo product, where the originally cis hydrogens are trans also. There are analogous anti additions among concerted cycloadditions. For example, heptafulvalene combines with tetracyanoethylene this way (Eq. 18). Heptafulvalene has 14 $\pi$

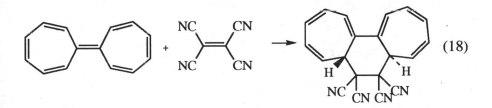

$$(18)$$

electrons, and the face-to-face reaction with an ethylene at the carbons shown is thermally forbidden, but by attaching at the top face at one end and the bottom at the other, the opposite phase interaction is reached and the reaction is thermally allowed. On the tetracyanoethylene side it is a syn reaction. Anti additions are called *antarafacial* and syn additions are called *suprafacial*. The process of Eq. 18 is abbreviated $[_\pi 14_a + _\pi 2_s]$ where the numbers indicate the number of electrons reorganizing in each piece and the subscripts a and s indicate antarafacial and suprafacial for each piece. Antarafacial cycloadditions occur in polyenes that are highly strained or twisted to allow such access.

The bracketed abbreviations may be applied to all the pericyclic reactions; that is, Eq. 10 is $[_\pi 4_s + _\pi 2_s]$, Eq. 12 is $[_\pi 2_s + _\pi 2_s]$, Eq. 15 is $[_\pi 6_s + _\pi 4_s]$, and Eq. 17 is $[_\pi 8_s + _\pi 2_s]$.

An "addition" to a $\sigma$ bond also offers the choice of antarafacial or suprafacial. Going back to an electrocyclic case, the thermal conrotatory opening of a cyclobutene has one face of a $\pi$ bond attaching to a front lobe from a $\sigma$ bond at one end and a back lobe at the other end. This is a $_\sigma 2_a$ process. Equation 19 indicates the new bonding with dashed lines.

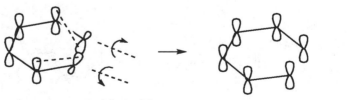

$$[_\sigma 2_a + _\pi 2_s] \tag{19}$$

The disrotatory opening of a cyclohexadiene has the front lobes on both ends (or back lobes on both ends) of the $\sigma$ bond attaching to the same face of the $\pi$ system, and this is a $_\sigma 2_s$ process (Eq. 20). In electrocyclic

$$[_\sigma 2_s + _\pi 4_s] \tag{20}$$

reactions there are two ways to abbreviate each process because the $\pi$ system could be considered antarafacially instead. Equation 19 could be abbreviated $[_\sigma2_s + _\pi2_a]$ and Eq. 20, $[_\sigma2_a + _\pi4_a]$. Both a and/or s designations are reversed, but the physical meaning remains the same.

Many photochemical intramolecular [2 + 2] cycloaddition reactions include breaking a $\sigma$ bond. In Eq. 21 we see that irradiation of a cy-

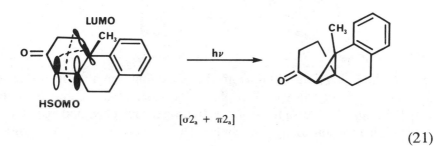

$$[_\sigma2_a + _\pi2_a]$$

$$(21)$$

clohexenone gives a three-membered ring. A $\sigma$ bond is broken from a stereogenic atom, but the "front" lobe of the $\sigma$ bond extending from the stereogenic atom was utilized in forming a new $\sigma$ bond, and the configuration is retained.[24] This is the expected result for a concerted reaction as indicated by the phase relationships for the $\pi$ HSOMO and the $\sigma$ LUMO. The new bonding is indicated by dashed lines. Look again at the breaking $\sigma$ bond. The $CH_2$ end of it is undergoing stereochemical inversion. Although this is not detectable here, it was found by means of deuterium labeling in a similar molecule.[25]

Whenever a $\sigma$ bond is broken in an antara fashion in a concerted reaction wherein both carbons are $sp^3$-hybridized at the finish (as in Eq. 21), one end will be stereochemically inverted and the other retained. If the $\sigma$ bond is broken in a supra fashion, both ends will be retained or both ends will be inverted.

The selection rules may now be summarized as follows. If the total number of electrons reorganizing in the two combining systems is $4n$ ($n$ = 1, 2, 3, . . .), [a + s] reactions are thermally allowed and [a + a] and [s + s] are photochemically allowed. If the total number of electrons is $4n + 2$, [s + s] and [a + a] are thermally allowed and [a + s] is photochemically allowed. The reverse of allowed reactions is also allowed.

### 8.3.3   Sigmatropic Reactions

In a sigmatropic reaction an allylic $\sigma$ bond cleaves and a new one forms further along the $\pi$ chain as exemplified in Eqs. 22 and 23. In Eq. 22

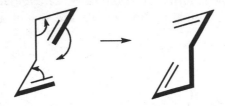

(22)

a carbon three positions along from the detaching one attaches to a site three positions along in the other chain. This is called a "shift of order [3,3]." In Eq. 23 the same atom, a hydrogen detaches and reattaches on the fifth position along the chain. This is called a "shift of order [1,5]."

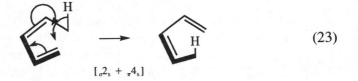

(23)

$$[_\sigma 2_s + _\pi 4_s]$$

For frontier orbital analysis this is viewed as a $\sigma$ bond combining with a diene bond. Examining the phase relationships of the diene HOMO with the sigma LUMO, we find an in phase overlap (Eq. 24) when the

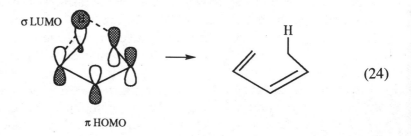

(24)

hydrogen reattaches to the same face of the $\pi$ system, that is, a suprafacial shift. This is a thermally allowed reaction. In contrast, a [1,3] suprafacial shift of a hydrogen is not allowed, as indicated by the antibonding relationship shown in Eq. 25.

$$\text{(25)}$$

If the [1,3] migrating group is a carbon instead of a hydrogen, there is a back lobe of opposite phase and the σ bond can be used antarafacially to give in phase overlap, but the migrating carbon will undergo inversion of configuration (Eq. 26). An example of this is the thermal rearrangement of the *exo*-bicyclo[2.1.1] compound in Eq. 27.[26] The new bonding

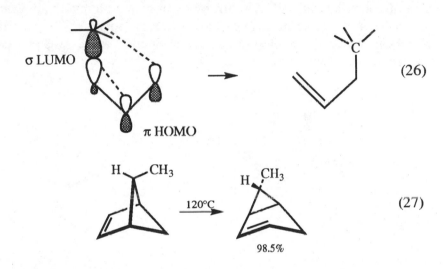

$$\text{(26)}$$

$$\text{(27)}$$

is indicated as dotted lines in Fig. XI. The endo isomer required a higher temperature and was not as cleanly in accord with the selection rules, but it gave mostly allowed products, including some migration of the $CH_2$ bridge (Eq. 28).

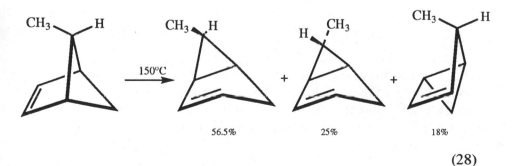

$$\text{(28)}$$

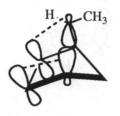

$$[\sigma 2_a + \pi 2_s] \text{ or } [\sigma 2_s + \pi 2_a]$$

**Figure XI.** A sigmatropic shift with inversion on carbon.

A concerted [1,5] shift of carbon is allowed suprafacially with retention of configuration. A [1,7] shift of hydrogen is thermally allowed if
it can reach the opposite face of the $\pi$ system (antarafacial) (Eq. 29).
A helical conformation allows this access. Such a process was observed
on heating the trienic ester in Eq. 30.[27]

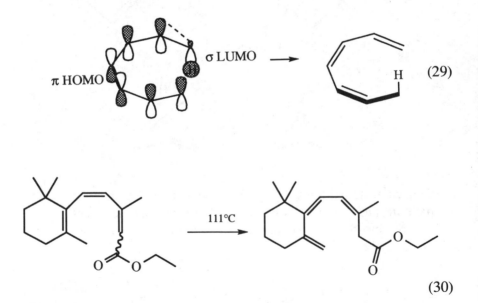

The [3,3] shift of Eq. 22 is actually a combination of three parts: a
$\sigma$ bond and two separate $\pi$ bonds. First a HOMO may be drawn that is
a combination of the $\sigma$ and one $\pi$ bond as in Eq. 31, and the closure to
a cyclic transition state is the addition of the second $\pi$ bond. The HOMO
involving four carbons has one node as in butadiene even though it is

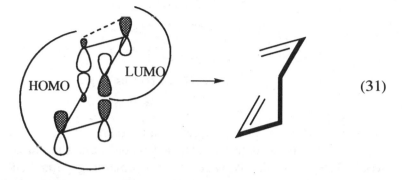

(31)

part $\sigma$ and part $\pi$.[28] This gives in-phase overlap with the ethylene LUMO, and the reaction is thermally allowed, suprafacial on both halves.

The selection rules for sigmatropic reactions of neutral molecules may be summarized as follows. If the number of electrons involved in reorganization (the sum of the numbers in brackets) is $4n + 2$, the reaction is thermally allowed suprafacially without inversion. The common cases are [1,5] and [3,3]. If the number of electrons is $4n$, the reaction is thermally allowed with an inversion or an antarafacial movement. The common cases are [1,3] and [1,7]. Remember that hydrogen cannot invert. The photochemical cases are the opposite of the thermal. Some structural features may make allowed reactions geometrically impossible.

There are more rules covering rearrangements of carbocations and carbanions beyond the scope of this book.[12]

### 8.3.4  Aromatic Stabilization of Transition States

In Section 8.2 aromatic character was connected with closed-shell electron occupation. In the usual cyclic $p$-orbital overlap there is a lowest-energy molecular orbital and above this, degenerate pairs of orbitals. Filling the lowest plus one or more degenerate pairs requires $4n + 2$ electrons. These are called *Hückel molecular orbitals*. As seen in Fig. VIII, $\psi_1$ of benzene has no nodal planes perpendicular to the plane of the ring, while $\psi_2$ and $\psi_3$ each have one nodal plane perpendicular to the ring. Such nodal planes intersect the ring in two places called *nodal zones*.

A Möbius strip is a band with a 180° twist in it such that the outside and inside surface are continuous and one (Fig. XII). Imagine opening a benzene ring, giving it a half twist and rejoining it. The lowest $\pi$

**Figure XII.** A Möbius strip.

molecular orbital would now have one nodal zone in it. Of the next-higher pair of molecular orbitals, one would gain a nodal zone and one would lose one, and likewise for the remaining pairs. The entire set would now be degenerate pairs (Fig. XIII), and a closed shell would require $4n$ electrons. Six electrons in Möbius benzene would give four bonding and two nonbonding and an incomplete shell.

Möbius cyclooctatetraene can now be neutral and aromatic (Fig. XIV) with eight electrons. Actually a twisted benzene or cyclooctatetraene ring would have very poor overlap from both lobes of each $p$ orbital. Anything smaller than a 22-membered ring would be poor.[29] However, special stability appears yet even when part of the ring involves only one lobe overlap.

Instead of ground-state molecules, let us now examine the cyclic transition states of several kinds of concerted reactions.[30,31] Reconsider the disrotatory electrocyclic closure of a hexatriene (Eq. 1). The transition state of this reaction (and the reverse) is shown in Fig. XVa, where a complete loop of overlap is followed with a dashed line from atom to atom through all six. The lower lobes could also be connected around part of the ring, but we will concentrate on one complete loop. This as drawn with a minimum of nodal zones resembles the lowest molecular orbital of aromatic benzene since there are six electrons and no nodes. This transition state is favored by aromatic stabilization and is reached only by disrotation. Conrotation would have given a transition state with one nodal zone (Fig. XVb) that would be part of a Möbius

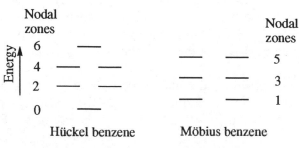

**Figure XIII.** Benzene $\pi$ molecular orbitals.

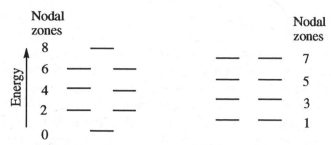

Nodal zones

Energy

8
6
4
2
0

Nodal zones

7
5
3
1

Hückel cyclooctatetraene    Möbius cyclooctatetraene

**Figure XIV.** Cyclooctatetraene $\pi$ molecular orbitals.

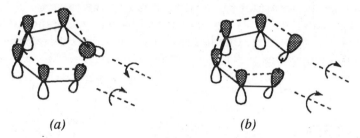

*(a)*                                    *(b)*

**Figure XV.** Transition states for closure of a hexatriene: (*a*) disrotatory, stabilized, allowed; (*b*) conrotatory, destabilized, forbidden.

orbital set, but six electrons do not give a closed shell here and the transition state does not have aromatic stabilization and is not allowed.

The transition state for the closure (or the reverse) of a butadiene is shown in Fig. XVI. The disrotatory closure allows a loop with no nodes (a Hückel orbital), but four electrons cannot give a closed-shell occupation (as with cyclobutadiene; Section 8.2) and the transition state is not stabilized (is forbidden). However, the conrotatory closure can be drawn with one nodal zone and therefore belongs to the Möbius group, where the four electrons will give a closed shell with aromatic stabilization; that is, this transition state is stabilized and allowed.

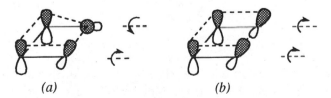

*(a)*                                    *(b)*

**Figure XVI.** Transition states for closure of a butadiene (*a*) disrotatory, destabilized, forbidden; (*b*) conrotatory, stabilized, allowed.

There are always two directions of conrotatory opening or closing to consider. In the opening of cyclobutenes there is a large electronic factor in substituents on carbons 3 and 4 that either favors outward rotation of the substituent, or inward rotation depending on the nature of the substituent.[32] Those that have nonbonding electron pairs are strongly inclined toward outward rotation, and those that have a low-lying vacant orbital that can accept electron density tend to rotate inward. There are numerous cases where this outweighs steric hindrance. In the opening of cyclobutenes, the transition state is early, the HOMO at that point is largely the breaking $\sigma$ orbital, and the LUMO is largely the $\sigma^*$ orbital. A substituent that has a low-lying vacant orbital (an acceptor substituent) can stabilize the transition state when it is rotating inward. This occurs by overlap with the HOMO to give an in-phase overlapping three-membered-ring transition state that has two electrons and is therefore aromatic (Eq. 32). The opening of 3-formylcyclobutene at 25–70°C, for example,

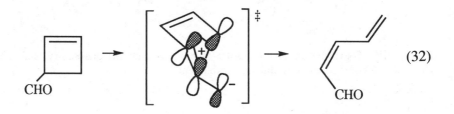

(32)

gives exclusively (Z)-pentadienal, the product of inward rotation of the formyl group.[33] A donor substituent would give the opposite effect (Eq. 33). It would provide a low-lying occupied orbital for overlap, and if it

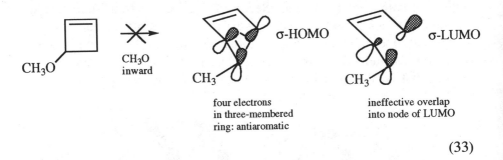

four electrons
in three-membered
ring: antiaromatic

ineffective overlap
into node of LUMO

(33)

rotated inward to overlap with the HOMO, it would give a four-electron antiaromatic three-membered ring, thus raising the activation energy for inward rotation. It could stabilize a LUMO, but inward rotation would give unfavorable overlap into a node in the LUMO. Outward rotation would be favored instead. Beyond the transition state, orbital mixing with the $\pi$ bond gives rapid stabilization for the conrotatory process. Equation 34 shows a reaction in which the outward preference of the donor methoxy group even overcomes the steric strain of a *tert*-butyl group rotating inward.[34]

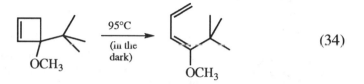

$$(34)$$

In Fig. XVII we see the transition states for a [4 + 2] and a [4 + 4] cycloaddition. The loop can be drawn with no nodes for each of them, meaning that they are Hückel-type orbitals where six electrons give closed shells [4 + 2] and a thermally allowed reaction, but eight [4 + 4] do not, and that reaction does not have an aromatic stabilized transition state and is not allowed.

In Fig. XVIII we see transition states for four possible sigmatropic shifts.

The advantage of this transition state analysis is that all types of concerted reactions are covered by basically one selection rule. If a continuous loop of overlap through all the atoms involved in the transition state can be drawn with no nodes, the reaction will be allowed thermally if it involves $4n + 2$ electrons. If the continuous loop requires one nodal zone, the reaction will be thermally allowed if it involves $4n$ electrons. Photochemical reactions are the opposite.

More strictly speaking, we are looking at the overlap of a basis set of atomic orbitals contributing to the molecular orbitals of the transition state. These may be drawn in any arbitrary phase orientation. In this way a Hückel set is recognized when strung together by the presence of an even number of nodal zones and a Möbius set is recognized by the presence of an odd number of nodal zones.

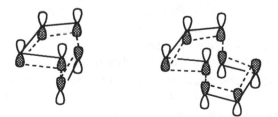

**Figure XVII.** Transition states for [4 + 2] and [4 + 4] cycloadditions.

[1,5] suprafacial shift of hydrogen, no nodes necessary, therefore Hückel; six electrons, therefore aromatic, allowed

[1,3] suprafacial shift of hydrogen, no nodes necessary, therefore Hückel; four electrons, therefore not aromatic, not allowed

[1,3] suprafacial shift of carbon with inversion, one nodal zone, therefore Möbius; four electrons, therefore aromatic, allowed

[3,3] suprafacial, suprafacial shift, no nodes, therefore Hückel; six electrons, therefore aromatic, allowed

**Figure XVIII.** Transition states for sigmatropic shifts.

Finally it must be kept in mind that although these theories have very broad predictive powers, there are many limitations. Other factors such as structural constraints can prevent a reaction that is allowed on an orbital basis. Which direction an allowed reaction will go is not predicted by this theory. Often more than one allowed reaction is possible from given starting materials, and again, these theories seldom make a distinction. Furthermore, nonconcerted mechanisms may operate to give products that are forbidden by these theories.

These reactions constitute sets of alternations rationalized with molecular orbitals. Regardless of whether you accept these theories, the experimental observations show a pattern of alternations, and that pattern alone can be used to make predictions.

## PROBLEMS

1. Calculate the delocalization energies of the allyl carbocation and the cyclopropenyl carbocation. Where is conjugation more effective? Do the same for the pentadienyl carbocation and the cyclopentadienyl carbocation. Where is conjugation more effective? Compare butadiene and cyclobutadiene this way. Compare hexatriene and benzene likewise. Is there a pattern to the results?

2. Three of the following structures were products from heating 2-chlorotropone with cyclopentadiene at 105°C. Which are the likely ones? Explain.[35]

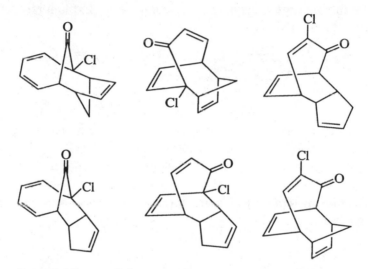

3. *trans*-3,4-Dichlorocyclobutene decomposes in 1.5 h at 90°C in dioxane, while the cis isomer requires 36 h at 180°C. Give the product(s) expected from each and explain the difference in thermal stability.[36]

4. Which isomer of 9,10-dihydronaphthalene is produced in each reaction? Explain.[37]

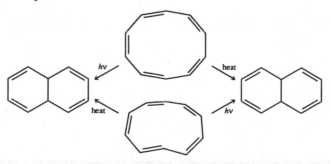

5. Heating the following acid–ester gave two acyclic products. Draw the likely structures.[38]

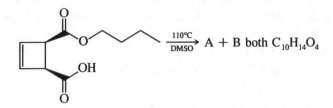

$110°C$ / DMSO → A + B both $C_{10}H_{14}O_4$

6. The following reaction occurred on standing in the dark at $25°C$.[39] Indicate the stereochemistry at the newly joined carbons; that is, are the hydrogens cis or trans to each other? Explain how you made your prediction.

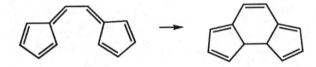

7. Although tropones give [4 + 2] adducts with maleic anhydride, the sulfur analog, cycloheptatrienethione, gives [8 + 2] cycloaddition.[40] Draw the structure of that product including stereochemical representation.

8. Explain in terms of one or more intermediates why the following reaction occurred:[41]

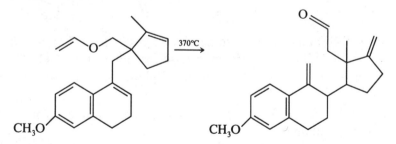

$370°C$ →

9. Draw the intermediate(s), and predict the complete stereochemistry of the product of the following reaction:[42]

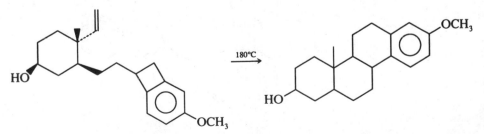

$180°C$ →

10. 1,3-Dithiepin gives the 2-deutero compound by exchange in deutero *tert*-butyl alcohol at 83°C catalyzed by potassium *tert*-butoxide. This happens over 150 times faster than the exchange on 1,3-dithiepane.[43] The difference was explained as a manifestation of aromatic stability in an intermediate. Is this reasonable? Explain why or why not.

11. Is the following reaction photochemical or thermal? Explain how you drew your conclusion.[44]

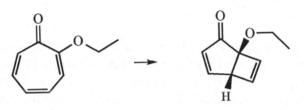

12. Heating the following compound gave an intramolecular [6 + 4] cycloaddition in 81% yield. Draw the structure of the product.[45]

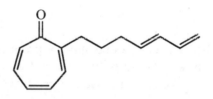

13. Give the expected stereochemistry (Z or E) about the double bonds in the product of the following reaction sequence [(a) reduction of the triple bond to a double bond with hydrogen over Lindlar's catalyst; (b) reduction of the lactone to a diol with diisobutylaluminum hydride at 0–25°C] involving reductions and a room-temperature rearrangement.[46] It is helpful to draw the transition state for the concerted step.

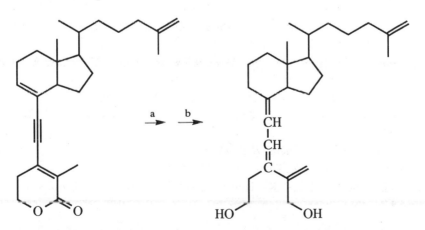

**14.** Propose a series of three pericyclic reactions that might explain the following result.[27]

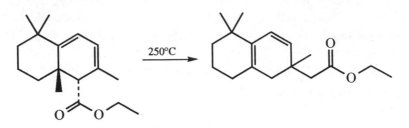

**15.** The following acyclic diol was treated with hydrogen over Lindlar's catalyst to give the bicyclic compound shown.[47] Draw the product, showing the stereochemistry at all four stereogenic carbons. Explain this stereochemical result by showing all the intermediates involved.

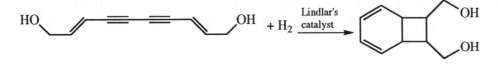

**16.** Heating the following compound gave a product isomeric with the starting material but showing five vinyl hydrogens in the NMR spectrum. What is the structure of the product?[48]

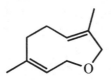

## REFERENCES

1. Yates, K. *Hückel Molecular Orbital Theory*, Academic Press, New York, 1978.
2. Zimmerman, H. E. *Quantum Mechanics for Organic Chemists*, Academic Press, New York, 1975.
3. Fukui, K. *Angew. Chem. Internatl. Ed.* **1982**, *21*, 801.
4. Breslow, R.; Höver, H.; Chang, H. W. *J. Am. Chem. Soc.* **1962**, *84*, 3168.
5. Breslow, R. *Acc. Chem. Res.* **1973**, *6*, 393.
6. Sondheimer, F. *Acc. Chem. Res.* **1972**, *5*, 81.
7. Swinborne-Sheldrake, R.; Herndon, W. C.; Gutman, I. *Tetrahedron Lett.* **1975**, 755.

8. Cook, M. J.; Katritzky, A. R.; Linda, P. *Advances in Heterocyclic Chem.* **1974,** *17,* 255–356.

9. Vogel, E.; Altenbach, H.-J.; Cremer, D. *Angew. Chem. Internatl. Ed.* **1972,** *11,* 935.

10. Goldberg, S. Z.; Raymond, K. N., Harmon, C. A.; Templeton, D. H. *J. Am. Chem. Soc.* **1974,** *96,* 1348.

11. *Pericyclic Reactions,* Vols. 1 and 2, Marchand, A. P.; Lehr, R. E., Eds., Academic Press, New York, 1977.

12. Woodward, R. B.; Hoffmann, R. *Angew. Chem. Internatl. Ed.* **1969,** *8,* 781.

13. Fukui, K. *Acc. Chem. Res.* **1971,** *4,* 57.

14. Fleming, I. *Frontier Orbitals and Organic Chemical Reactions,* Wiley-Interscience, Chichester (UK), 1976.

15. Marvell, E. N. *Thermal Electrocyclic Reactions,* Academic Press, New York, 1980.

16. Marvell, E. N.; Caple, G.; Schatz, B.; Pippin, W. *Tetrahedron* **1973,** *29,* 3781.

17. Winter, R. E. K. *Tetrahedron Lett.* **1965,** 1207.

18. Doorakian, G. A.; Freedman, H. H. *J. Am. Chem. Soc.* **1968,** *90,* 5310, 6896.

19. Courot, P.; Rumin, R. *Tetrahedron Lett.* **1970,** 1849.

20. Herndon, W. C. *Chem. Rev.* **1972,** *72,* 157–179.

21. Houk, K. N. In *Pericyclic Reactions,* Vol. II, Marchand, A. P.; Lehr, R. E., Eds., Academic Press, New York, 1977, Chapter 4.

22. Saito, K.; Ida, S.; Mukai, T. *Bull. Chem. Soc. Jpn.* **1984,** *57,* 3483.

23. Farrant, G. C.; Feldman, R. *Tetrahedron Lett.* **1970,** 4979.

24. Chapman, O. L.; Sieja, J. B.; Welstead, Jr., W. J. *J. Am. Chem. Soc.* **1966,** *88,* 161.

25. Bellus, D.; Kearns, D. R.; Schaffner, K. *Helv. Chim. Acta.* **1969,** *52,* 971.

26. Roth, W.; Friedrich, A. *Tetrahedron Lett.* **1969,** 2607.

27. Fráter, G. *Helv. Chim. Acta* **1974,** *57,* 2446.

28. Fleming, I., ref. 14, p. 102.

29. Heilbronner, E. *Tetrahedron Lett.* **1964,** 1923.

30. Dewar, M. J. S. *Tetrahedron Supplement* **1966,** *8,* 75.

31. Zimmerman, H. E. In *Pericyclic Reactions,* Vol. 1, Marchand, A. P.; Lehr, R. E., Eds., Academic Press, New York, 1977, Chapter 2.

32. Rondan, N. G.; Houk, K. N. *J. Am. Chem. Soc.* **1985,** *107,* 2099.

33. Rudolf, K.; Spellmeyer, D. C.; Houk, K. N. *J. Org. Chem.* **1987,** *52,* 3708.

34. Houk, K. N.; Spellmeyer, D. C.; Jefford, C. W.; Rimbault, C. G.; Wang, Y.; Miller, R. D. *J. Org. Chem.* **1988,** *53,* 2125.

35. Ito, S.; Sakan, K.; Fujise, Y. *Tetrahedron Lett.* **1969,** 775.

36. Müller, P.; Joly, D.; Mermoud, F. *Helv. Chim. Acta* **1984,** *67*, 105.

37. Masamune, S.; Hojo, K.; Bigam, G.; Rabenstein, D. L. *J. Am. Chem. Soc.* **1971,** *93*, 4966.

38. Trost, B. M.; McDougal, P. G. *J. Org. Chem.* **1984,** *49*, 458.

39. Sauter, H.; Gallenkamp, B.; Prinzbach, H. *Chem. Ber.* **1977,** *110*, 1382.

40. Machiguchi, T.; Hoshino, M.; Ebine, S.; Kitahara, Y. *J. Chem. Soc. Chem. Commun.* **1973,** 196.

41. Ziegler, F. E.; Lim, H. *J. Org. Chem.* **1982,** *47*, 5229.

42. Kametani, T.; Suzuki, K.; Nemoto, H. *J. Chem. Soc. Chem. Commun.* **1979,** 1127.

43. Bordwell, F. G.; Fried, H. E. *J. Org. Chem.* **1991,** *56*, 4218.

44. Kaftory, M.; Yagi, M.; Tanaka, K.; Toda, F. *J. Org. Chem.* **1988,** *53*, 4391.

45. Rigby, J. H.; Moore, T. L.; Rege, S. *J. Org. Chem.* **1986,** *51*, 2398.

46. Barrack, S. A.; Gibbs, R. A.; Okamura, W. H. *J. Org. Chem.* **1988,** *53*, 1793.

47. Lindberg, T., Ed., *Strategies and Tactics in Organic Synthesis*, Academic Press, New York, 1984, pp. 160–164.

48. Marshall, J. A.; Lebreton, J. *J. Org. Chem.* **1988,** *53*, 4108.

# 9

# PHYSICAL INFLUENCES ON REACTIONS

The experimental section of a journal article is the core of factual observation. It remains of value even when the explanations and theories are revised. This is the part of the article that is consulted in detail when you want to prepare a reported compound or a closely related one. From introductory texts it is easy to get the impression that simply mixing starting materials gives the product. Some cases are as easy as that, but many require particular conditions or techniques to stimulate or control the process.*

The conditions are chosen for efficiency and selectivity. The reactants may be capable of giving many products, but with appropriate handling the desired one may predominate. Appropriate techniques may make the reaction easy to conduct, without temperature extremes or lengthy reaction times. Many reactions are kinetically controlled, that is, the reactants given sufficient energy and opportunity will proceed to products. Other reactions are thermodynamically controlled, and will proceed readily in either direction until an equilibrium amount of reactants and products are present. In the latter, progress toward products is made by using an excess of a reactant or removing a product from the reaction phase by distillation or crystallization.

The larger part of the effort is usually the separation and purification of the products, but these procedures are found in laboratory texts and will not be covered here.

*There are many safety considerations in the design of experiments, and this should be done under the supervision of experienced personnel.

## 9.1  UNIMOLECULAR REACTIONS

Many eliminations and rearrangements require only heat to proceed, and the simplest procedure is heating with no solvent. Heating dicyclopentadiene at 200° to 210°C gives the retro-Diels–Alder product cyclopentadiene, which escapes as a gas and is condensed.[1] Heating a chlorooxazolidinone neat gives HCl elimination directly (Eq. 1).[2]

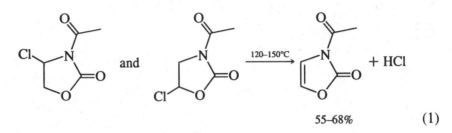

55–68%                    (1)

More often reactions are carried out in a solvent, which may serve several purposes. A solvent with an appropriate boiling point may be heated at reflux to provide a constant temperature for the reaction. If it is particularly exothermal, the solvent can be a heat sink to dissipate the exotherm rather than allowing the reactant to rise precipitously to a high temperature. If there is a competing bimolecular reaction between molecules of the reactant, the solvent is a diluent that lessens the collisions and favors the unimolecular reaction.

An alternative that avoids the expense of a solvent and its removal is gas-phase flow thermolysis.[3] This avoids bimolecular reactions because in the low density of a gas (especially under vacuum) molecular collisions are relatively infrequent. Short reaction times at extremely high temperatures may be used where warranted. For example, cis-4b,8a-dihydrophenanthrene was prepared in high yield by passing the benzotricyclic compound (Eq. 2) through a tube at 1 mm pressure and 550°C, even though the product disproportionates (bimolecular reaction) in the liquid state at 150°C.[4]

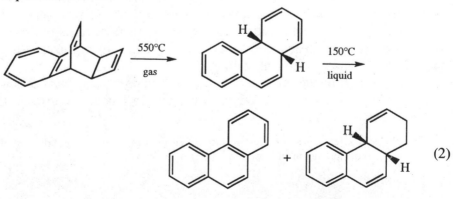

(2)

## 9.2  HOMOGENEOUS TWO-COMPONENT REACTIONS

Efficient interaction of two or more components occurs if they are combined in a single liquid phase. Cyclooctatetraene and maleic anhydride may be heated together at 165–170°C for 30 min to give a high yield of the Diels–Alder adduct.[5]

If one or more of the reactants remain solid at the desired temperature, the reaction zone is often limited to the surface of the solid and may soon be blocked by a layer of reaction product. A solvent that dissolves both reactants can alleviate this problem and also aid in temperature control. More solvent will give less frequent collisions between reactants and slow the progress. This may be desirable for moderation of very fast reactions or minimized for slower ones. In some cases one reactant may be used in excess as the solvent to maintain the high collision frequency.

It is common to control fast reactions by adding one of the reactants over a period of time to a solution of the other(s). Under these conditions, the added component does not become diluted by all the solvent in the reaction flask, and local high concentrations exist. This is usually of no consequence, but with especially reactive reagents, competing reactions may occur in those concentrated zones. For example, when the reaction in Eq. 3 was carried out by adding the 1-aminobenzotriazole

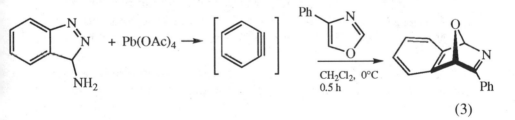

(3)

(ABT) to a solution of lead tetraacetate and 4-phenyloxazole, the benzyne intermediate was generated in local high concentration where it dimerized to biphenylene, and the Diels–Alder product was obtained in low yield. This is corrected by adding the ABT and lead tetraacetate simultaneously on opposite sides of the stirred solution. In this way there is no local high concentration of benzyne and the Diels–Alder adduct is obtained in quantitative yield.[6] It is interesting to note that if the product is heated in benzene, a retro-Diels–Alder reaction occurs affording isobenzofuran (a potent diene for other Diels–Alder reactions) along with benzonitrile.

Another example was found in the attempted monoacylation of diamines with benzoyl chloride. Addition of a solution of 2 mmol of benzoyl chloride in $CH_2Cl_2$ to a vigorously stirred solution of 10 mmol of

1,2-ethanediamine in $CH_2Cl_2$ at $-78°C$ gave 0.99 mmol of the diacylated product. The high concentration of the acid chloride at the contact site gave rapid diacylation. Using the slower reacting benzoic anhydride instead of the acid chloride lowered the diacylation yield to 0.24 mol. Greatly diluting the anhydride solution lowered it further to 0.14 mmol. A statistical yield would be 0.10 mmol.[7]

## 9.3    TEMPERATURE OPTIMIZATION

The temperature selected for a reaction is a compromise, arrived at by experimentation using similar cases as guides. A lower temperature will give a longer reaction time, while a higher temperature will bring on competing reactions that will lessen the yield of the desired product and complicate the purification process.

Careful selection of temperatures is necessary in the coupling reactions of alkyl cuprates with primary halides (Eq. 4). The cuprates are

$$RLi + CuI \xrightarrow{\text{THF}} R_2CuLi \xrightarrow{\text{RX}} R—R' + RCu + LiX \qquad (4)$$

thermally unstable and are prepared at low temperatures and used promptly. Starting with $n$-butyllithium the sequence can be done at $-78°C$, but little coupling is found with $sec$-butyl and $tert$-butyllithium. If a temperature of $0°C$ is used briefly for the first step and then the coupling done at $-78°C$, good yields are found with all three lithium reagents.[8,9]

The reaction of LDA with carboxylic acids to afford the $O,\alpha$-dianions proceeds rapidly at $-40°C$, but if the carbon chain is very long, the reaction is very slow. Eicosanoic acid gave only the monoanion at room temperature, but heating at $50°C$ for several hours gave the dianion.[10] Similar effects are found in other reactions of long-chain molecules.

## 9.4    PRESSURE EFFECTS

Laboratory reactors are available from many suppliers for applying pressures of up to 20 kbar (19,700 atm, 2 GPa) to reaction solutions in volumes of 1–50 mL. Pressures above 5 kbar can have a marked influence on reactions.[11,12] Those reactions that have negative activation volumes ($\Delta V^{\ddagger}$); that is, those in which the transition state including solvation occupies a smaller volume than the starting molecules, will be

considerably accelerated by very high pressures. For example, the esterification of acetic acid with ethyl alcohol at 50°C is five times faster
at 2 kbar (compared to 1 atm), and 26 times faster at 4 kbar.[13]

Negative $\Delta V^{\ddagger}$ values are common among quaternization of amines
and phosphines, hydrolyses and esterifications, Claisen and Cope rearrangements, nucleophilic substitutions, and Diels–Alder reactions.

Very high pressure is especially valuable where heat alone leads to
alternate reaction products. For example, the phosphonium salt in Eq.
5 was prepared in excellent yield at 20°C and 15 kbar. At 1 atm and

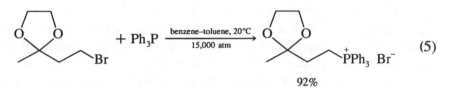

$$(5)$$

92%

20°C no reaction occurs, and at 80°C only decomposition products are
found.[14] Diels–Alder reactions may be aided by high pressure, also,[15]
as in the case shown in Eq. 6. Heating the diene and dienophile in
toluene at reflux at 1 atm for 65 h gave some recovered starting materials
and unidentified products, but none of the desired adduct. However, 2
days under 10 kbar pressure gave the Diels–Alder product in 90% yield.

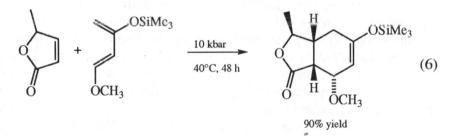

$$(6)$$

90% yield

Esters may be hydrolyzed under mildly basic conditions at room temperature at 10 kbar. This avoids the potential side reactions, such as
epimerization, dehydration, and migration of double bonds, which occur
in hot acidic or basic conditions at 1 atm on sensitive compounds like
that in Eq. 7.[16]

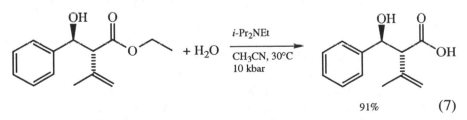

91% $$(7)$$

Besides accelerating reactions, high pressure may be used to shift the position of an equilibrium. At 100°C and 1 atm the Diels–Alder reaction of naphthalene and excess maleic anhydride reaches equilibrium at only 1% conversion. At 100°C and 10 kbar the yield is 80%.[17]

## 9.5  POLAR SOLVENTS

It is frequently necessary to treat an organic compound with an ionic reagent. The ionic reagents are not appreciably soluble in nonpolar organic solvents; therefore, polar solvents are used.[18] The ions are solvated by coordination with the oppositely charged end of the solvent dipole or by specific hydrogen bonding. The hydrogen bonding (protic) solvents give higher rates in $S_N1$ reactions compared to aprotic solvents because they aid the departure of anionic leaving groups by hydrogen bonding with them. The more frequently used $S_N2$ reactions are generally not aided by these protic solvents because they cluster around the nucleophile anion, rendering it less reactive. The smaller the anion, the more concentrated the charge and the more tightly it will be solvated by protic solvents. Thus in methanol the order of nucleophilicity of halides toward iodomethane is $I^- > Br^- > Cl^- > F^-$.[19]

The $S_N2$ reactions are greatly aided by dipolar aprotic solvents such as DMSO, DMF, hexamethylphosphoric triamide, tetramethylurea, or 1,3-dimethyl-2-imidazolidinone. In these solvents the positive end of the dipole is relatively encumbered while the negative end is exposed and available for association with cations. The anions are thus relatively little solvated and exceptionally reactive. For example, NaCN in DMSO reacts with primary and secondary alkyl chlorides to give nitriles in 0.5–2 h, while 1–4 days are required in aqueous alcohol.[20,21] In these dipolar aprotic solvents the relative nucleophilicity of anions follows charge density and is the reverse of that found in protic solvents. The displacement on $n$-butyl tosylate in DMSO gives the order $F^- > Cl^- > Br^- > I^-$.[22]

An empirical scale of solvent polarity has been developed on the basis of shifts of a UV–visible absorption maximum of a pyridinophenylate indicator,[23] which should be useful for predicting solvent effects on rates.

## 9.6  REACTIONS WITH TWO LIQUID PHASES

The water solubility, low volatility, and reactivity of the dipolar aprotic solvents can be disadvantageous; therefore, another approach is of value. Many inorganic reagents are soluble in water and are used with an organic solution, with vigorous stirring to promote a reaction at the surface between the water and organic phases. However, the frequency of successful collision on a surface is far less than in the bulk of a homogeneous solution. Fortunately, in a great many cases this difficulty is readily overcome by adding a small amount of a tetraalkylammonium or phosphonium salt. The quaternary cations with sufficiently large alkyl groups have an affinity for organic solvents and will carry reactive anions with them into solution in the organic layer. These anions are particularly reactive because they carry only a small hydration shell. Some stirring is still necessary because the quaternary salts are used in catalytic amounts and must repeatedly exchange product anions for reactant anions at the phase boundary. This is called *phase-transfer catalysis*.[24-27]

Another mode of operation of these catalysts occurs in the formation of carbanions using concentrated aqueous NaOH. A proton is removed from a precursor by the $OH^-$ ion at the interface, and the carbanion moves into the bulk of the organic phase accompanied by the lipophilic cation. This allows formation of substantial concentrations of carbanions that are ordinarily more basic than aqueous $OH^-$ and yet avoids the difficulties of using stronger bases such as sodamide or LDA under anhydrous conditions.

Similar effects are found with the more expensive macrocyclic polyethers (crown ethers), which complex alkali metal cations, giving them sufficient lipophilicity to dissolve in the organic phase, bringing along the reactive anion.

Alkyl halides and sulfonates will undergo nucleophilic substitution by these reactive inorganic anions or carbanions in the organic layer. The product halide ions are nucleophilic and may compete with reactant anions that have leaving-group ability and reach an equilibrium condition that is dependent largely on the relative amounts of the two anions in the organic phase. Chloride or bromide ions cannot displace iodide or tosylate ions efficiently because the iodide and tosylate anions are relatively lipophilic and remain in the organic phase. One can convert organochloride to bromide or iodide and convert organobromide to iodide in good yield. Mesylates are good leaving groups with low lipophilicity and are displaceable by all halides (Eq. 8).[28] Other nucleophilic

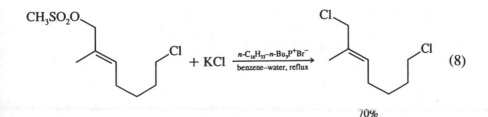

$$CH_3SO_2O \overset{Cl}{\diagup} + KCl \xrightarrow[\text{benzene-water, reflux}]{n\text{-}C_{16}H_{33}\text{-}n\text{-}Bu_3P^+Br^-} Cl \overset{Cl}{\diagup} \quad (8)$$

70%

ions that are not themselves good leaving groups, such as cyanide, phenoxides, carboxylates, carbanions, alkoxides (Chapter 4, Eq. 39), or sulfinates (Eq. 9)[29] can give high yields of substitution products.

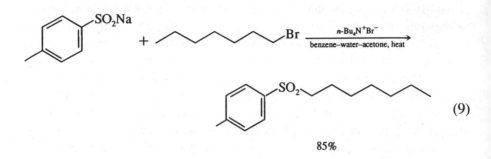

$$(9)$$

85%

Oxidation reactions using permanganate[30, 31] (Section 4.1.1), dichromate, or hypochlorite are very effective with phase-transfer catalysis. Borohydride reductions are facilitated, also.[32] Dihalocarbenes can be made using aqueous NaOH with phase-transfer catalysis.[33]

## 9.7    REACTION OF A SOLID WITH A LIQUID

If an ionic inorganic reagent is combined with a low-polarity organic solution without a water layer, insolubility is still a problem. An obvious resort is to use a more organic soluble salt of the same reactive anion. For example, although sodium borohydride has very low solubility in dichloromethane, ether, or THF, tetrabutylammonium borohydride has high solubility in methylene chloride. This salt is useful for the reduction of aldehydes and ketones that are not soluble in hydroxylic solvents or that react with those solvents to give hydrates, acetals, or ethers.[34]

Potassium permanganate is insoluble in organic solvents, and when used in water, it decomposes, requiring a large excess. Tetrabutylammonium permanganate is easily prepared and may be used in stoichiometric amount in pyridine solution at room temperature to give oxidation products quickly and in high yield.[35]

The quaternary salts can be used in catalytic amount in truly heterogeneous conditions. For example, solid potassium phthalimide was used in toluene in a Gabriel synthesis of amines:[36]

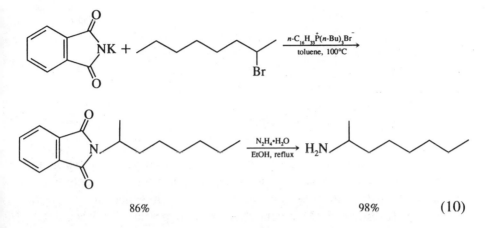

86% 98% (10)

One may even omit the solvent altogether. Mesitoic esters are slow to form and slow to hydrolyze under ordinary conditions, owing to steric hindrance. Using solid KOH and methyltrioctylammonium chloride (Aliquat 336), this can be done with the organic reactants and products serving as the liquid phase (Eq. 11).[37] When the hydrolysis is carried

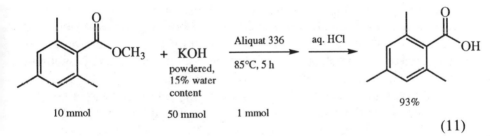

10 mmol          50 mmol     1 mmol                (11)

out with water and hydrocarbon liquid phases, yields are low. The esterifications were carried out by mixing the acid, KOH, alkyl bromide or iodide, and Aliquat 336 at 20–85°C to afford mesitoates in 88–98% yield.

Alkali metal salts become soluble in low-polarity organic solvents when those cations are coordinated by close-fitting cyclic ethers, giving them an organic compatible exterior.[38] Dicyclohexyl-18-crown-6 is particularly suitable for potassium, and the combination of it and potassium permanganate has a high solubility in benzene.

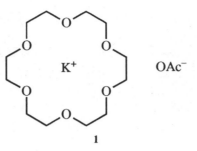

**1**

Substitution reactions may be carried out in high yield by using 18-crown-6 with solid potassium salts. Acetate anion has low nucleophilic activity ordinarily, but potassium acetate in the presence of a catalytic amount of 18-crown-6 (**1**) reacts readily with primary and secondary alkyl bromides and also tertiary chlorides to give alkyl acetates (Eq. 12).[39] This was run under heterogeneous conditions, with excess potas-

$$n\text{-}C_6H_{13}Br + KOAc \xrightarrow[\substack{CH_3CN, 83°C \\ 3\,h}]{\text{18-crown-6}} n\text{-}C_6H_{13}OAc + KBr \quad (12)$$

100%

sium acetate present as a solid. The solubility of potassium acetate in acetonitrile at 25°C is only $5 \times 10^{-4}$ $M$, but with 0.14 $M$ 18-crown-6, the solubility is 0.1 $M$. In acetonitrile or benzene with the crown ether, the acetate is not encumbered by a hydrogen-bonding solvent shell, and it is not closely associated with the potassium cation; therefore, it is quite reactive as a nucleophile. Under the conditions of Eq. 12 without the crown ether, essentially no reaction occurred.

Solid potassium cyanide reacts with primary and secondary chlorides in acetonitrile or benzene solution with vigorous stirring in the presence of 18-crown-6.[40] The term "phase-transfer catalysis" is applied to these solid–liquid reactions, as well as to the liquid–liquid cases in Section 9.6.

In the preceding cases the intent is to bring at least small amounts of the reagent anions into the organic solution for reaction. In contrast to these, solid metals must undergo oxidation at their surfaces, and obtaining convenient rates of reaction depends on selecting appropriately fine particles. Zinc metal is available as lumps (mossy), granular particles, and as dust. Even finer zinc may be prepared by reducing anhydrous zinc chloride with potassium metal in THF. This form is sufficiently reactive to convert alkyl bromides to dialkylzinc compounds.[41]

Magnesium in the form of lathe turnings is sufficient for most Grignard reactions, but for some temperature-sensitive or reluctant cases, activation is needed. Stirring magnesium turnings under a nitrogen atmosphere with a Teflon bar for 15 h gives smaller dark-gray particles. This reacts promptly with allylic and benzylic chlorides in ether at 0°C to give clear solutions of the Grignard reagents in high yield, with no coupling products. With untreated turnings, the slower-reacting magnesium allows a competing reaction of the halide with the Grignard reagent as it is formed, leading to appreciable amounts of coupled hydrocarbon and precipitated $MgCl_2$.[42] A fine powder from potassium metal reduction of $MgCl_2$ is even more reactive.[41]

Sodium is available as cast ingots that can be cut into small pieces with a knife. When more surface area is needed, it can be finely pulverized by stirring it molten in refluxing toluene and then cooling. This was done for the process in Eq. 13,[43] where the cooled toluene was replaced with ether. Trimethylchlorosilane is slow to react with carbanions; therefore, it can be present in a reaction mixture where it will capture oxyanions as they are formed.

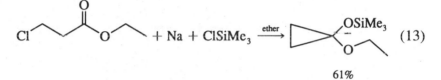

$$+ Na + ClSiMe_3 \xrightarrow{\text{ether}} \qquad (13)$$

61%

Metals can be made to react dramatically faster using ultrasound (sonication).[44-46] A reaction flask may be partly immersed in an ultrasonic bath, or an ultrasonic probe can be inserted directly in a reaction mixture. The oscillator is typically 60–250 W at 20–60 kHz. A mixture of reactive metal, organohalide, and solvent may be ultrasonically irradiated to give an organometallic reagent. The cavitation effect of the ultrasound causes rapid erosion of the metal surface. Electrophiles may often be included in the original reaction mixture to react with the organometallic reagent as it is formed. The Barbier reaction with lithium metal is an example of this:[47]

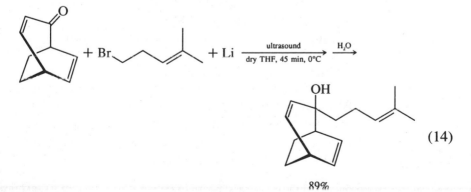

$$+ Br \qquad + Li \xrightarrow[\text{dry THF, 45 min, 0°C}]{\text{ultrasound}} \xrightarrow{H_2O} \qquad (14)$$

89%

The Reformatsky reaction, in the presence of iodine, proceeds in high yield in minutes at 25–30°C with ultrasonic erosion of ordinary zinc dust in dioxane (but not in ether or benzene).[48]

Solid inorganic reagents will react with solutions of organic compounds under ultrasonic irradiation. Dry powdered potassium permanganate is insoluble in hexane and gives only a small amount of oxidation when stirred with a solution of 2-octanol in hexane. However, in an ultrasonic bath, 2-octanone was obtained in 93% yield after 5 h. Several other secondary alcohols were oxidized likewise in good yield in benzene.[49]

Sodium hydroxide is insoluble in chloroform, but with ultrasound it reacts rapidly to give dichlorocarbene. Dry powdered sodium hydroxide stirred in a chloroform solution of an alkene immersed in an ultrasonic bath affords dichlorocyclopropanes in high yields:[50]

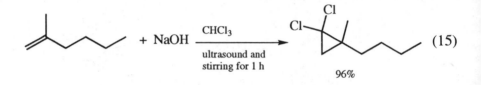

$$\qquad\qquad (15)$$

## 9.8   REACTIONS ON INORGANIC SOLID SUPPORTS

Numerous reactions that are inefficient in solution will proceed at lower temperatures and with higher selectivity in the adsorbed state on porous inorganic solids.[51,52] Just as polar solvents have substantial effects on reactions in homogeneous solution, the very polar adsorbants can affect reactants. They modify the dipolar character of the molecules, or they may hold molecules in a reactive orientation in pores in the solid. In some cases a solvent is present to deliver and remove molecules to and from the solid surface, and in other cases no solvent is present at all.

Ozone, because of its very low solubility in ordinary organic solvents, is suitable only for very fast reactions such as with alkenes. Benzene rings react at $10^{-5}$ times the rate of alkenes and are not conveniently ozonolyzed in solution. Silica gel will adsorb up to 4.7% by weight of ozone at −78°C. If an organic substrate is first adsorbed on the silica gel and then ozone added at −78°C, followed by warming to room temperature, oxidation of benzene rings, and also tertiary hydrogen sites, occurs conveniently and in good yield (Eqs. 16,[53] 17[54]). Adamantane was converted to 1-adamantanol similarly in 81–84% yield.[55] Amines are cleanly converted to nitro compounds similarly.[56]

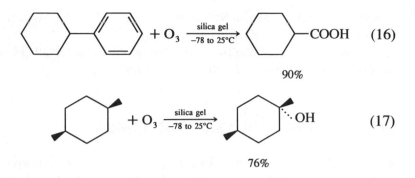

$$90\%$$

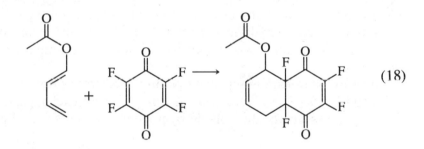

$$76\%$$

Nitroalkanes are converted to ketones and aldehydes by a variety of methods in solution involving strong acids or redox reactions. A very simple alternative is to adsorb the material on silica gel rendered basic with sodium methoxide. Elution with ether then gives the carbonyl compound in high yield and purity.[57]

The Diels–Alder reaction shown in Eq. 18 requires heating at 96°C for several hours, but on silica gel it occurs at room temperature.[58]

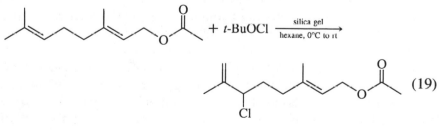

The chlorination of alkenes with *tert*-butyl hypochlorite on silica gel is clean, fast, and selective,[59] as exemplified in Eq. 19.[60] In the absence of silica gel, no reaction occurs under these conditions. Ethers are oxidatively cleaved to aldehydes and ketones in high yields by heating in

$$82\%$$

isooctane or $CCl_4$ with $Cu(NO_3)_2$ on silica gel. Other powders such as zeolites, cellite, and carbon were ineffective.[61]

A great many other reactions are promoted by solid adsorbants, including silica gel, alumina, clay, and carbon.

## 9.9  USING UNFAVORABLE EQUILIBRIA

Most of the techniques covered thus far are designed to improve contact among reactants where reactions would otherwise be too slow (kinetically limited). Some reactions are fast in forward and reverse directions, leading to small conversions at equilibrium (thermodynamically limited). Many of these may be pushed toward high conversions by a phase change. If one of the products of an unfavorable equilibrium is removed from the reaction solution by evaporation or crystallization while the reaction is under way, nearly all the starting material may be converted. The equilibrium is continually reestablished based on remaining concentrations. Distillation is used to remove a lower-boiling product, as in trans ketalization, trans esterification, preparation of higher anhydrides from acetic anhydride (Chapter 4, Eq. 15), and many others. Azeotropic distillation of water from esterifications or enamine syntheses is common.

Ordinarily, sulfonate groups may be displaced from primary carbons by bromide ions. The reverse can be done with an exchange reaction driven by the escape of the volatile byproduct bromomethane, bp 4°C:

$$Cl\diagdown\diagup{}_{Br} + CH_3OTs \underset{135°C, 2\,h}{\overset{Bu_4N^+Br^-}{\rightleftharpoons}} Cl\diagdown\diagup{}_{OTs} + CH_3Br \qquad (20)$$

$$68\%$$

One component of an equilibrium may have a lower solubility than another and thus may separate as formed allowing gradual conversion of nearly all to the low solubility component, while the relative amounts in solution remain near equilibrium. For example, the preparation of the nitrone in Eq. 21 reaches an equilibrium yield of 70% when equimolar amounts of aldehyde and hydroxylamine are dissolved in ethanol. Since water is a product of the reaction, adding more water to the solution should shift the equilibrium back toward more reactants *in solution*. However, the product is insoluble in the water–alcohol, and it precipitates. The mobile equilibrium provides more product, which continues to come out of solution until reaching an isolated yield of 85%.[62]

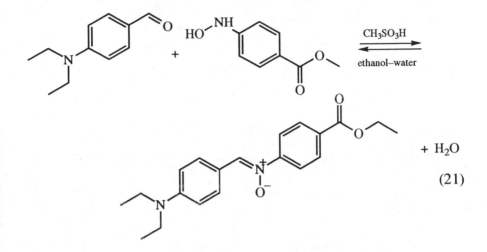

$$+ H_2O$$

(21)

Equilibria among stereoisomers have been manipulated using solubility differences. Meso and racemic dibromoglutaric acids are present in equal amounts when equilibrated in the presence of HBr (Eq. 22).

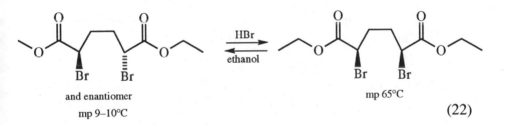

and enantiomer
mp 9–10°C

mp 65°C

(22)

The higher-melting meso compound is less soluble in ethanol, and will crystallize from the equilibrium solution when cooled to 5–10°C. Filtering three crops of crystals, concentrating the solution after each, converted 82% of the mixture to the meso isomer.[63] In other examples, an enantiopure compound was converted to an enantiopure epimer by equilibration where the product solidified.[64, 65]

Finally, a racemic amine was converted to one pure enantiomer in 91% yield by equilibrating the enantiomers in the presence of (+)-10-camphorsulfonic acid, which preferentially crystallized as the salt of the S enantiomer:[66]

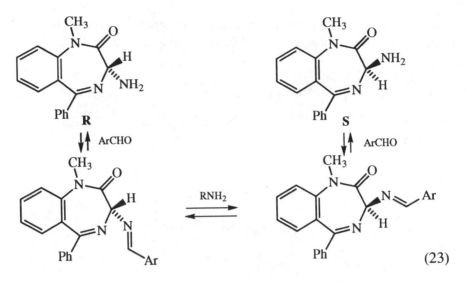

A catalytic amount of an aromatic aldehyde gave an imine that could racemize in solution in isopropyl acetate–acetonitrile while the salt of the $S$ isomer crystallized.

## PROBLEMS

1. The Grignard reagents from 2-(2-bromoethyl)-1,3-dioxolane and 2-(3-chloropropyl)-1,3-dioxolane are thermally unstable and decompose during their preparation under ordinary conditions.[67,68] What can be done to overcome this problem?[69]

2. Ozone was passed into a solution of 26 g of *cis*-decalin in $CCl_4$ at 0°C for 147 h. This gave 7 g of decahydronaphth-4a-ol.[70] What could be done to improve the yield and shorten the reaction time?

3. Heating 1-phenyl-1-benzoylcyclopropane with excess ethyl bromo-acetate and 20-mesh zinc at reflux for 17 h gave a 47% yield of the Reformatsky product along with 55% recovered ketone.[71] What could be done to improve this reaction?

4. The oxidation of alkynes to $\alpha$-diketones with potassium permanganate is a well-known reaction, but the early examples are carboxylate salts that are soluble in aqueous permanganate. What conditions would you choose to oxidize 1-phenyl-1-pentyne to 1-phenyl-1,2-pentanedione?[72]

# REFERENCES

1. Partridge, J.; Chadha, N. K.; Uskovic, M. R. *Org. Synth.* **1985,** *63*, 44.

2. Scholz, K.-H.; Heine, H.-G.; Hartmann, W. *Org. Synth.* **1984,** *62*, 149.

3. Karpf, M. *Angew. Chem. Internatl. Ed.* **1986,** *25*, 414–430.

4. Paquette, L. A.; Kukla, M. J.; Stowell, J. C. *J. Am. Chem. Soc.* **1972,** *94*, 4920.

5. Reppe, W.; Schlichting, O.; Klager, K.; Toepel, T. *Annalen Chem.* **1948,** *560*, 66.

6. Whitney, S. E.; Rickborn, B. *J. Org. Chem.* **1988,** *53*, 5595.

7. Jacobson, A. R.; Makris, A. N.; Sayre, L. M. *J. Org. Chem.* **1987,** *52*, 2592.

8. Schwartz, R. H.; San Filippo, Jr., J. *J. Org. Chem.* **1979,** *44*, 2705.

9. Posner, G. H. *Org. React.* **1975,** *22*, 253.

10. Belletire, J. L.; Fry, D. F. *J. Org. Chem.* **1987,** *52*, 2549.

11. Le Noble, W. J., Ed. *High Pressure Organic Chemistry*, Elsevier, Amsterdam, 1988.

12. Matsumoto, K.; Acheson, R. M., Eds. *Organic Synthesis at High Pressures*, Wiley-Interscience, New York, 1991.

13. P'eng, S.; Sapiro, R. H.; Linstead, R. P.; Newitt, D. M. *J. Chem. Soc.* **1938,** 784.

14. Dauben, W. G.; Gerdes, J. M.; Bunce, R. A. *J. Org. Chem.* **1984,** *49*, 4293.

15. Ortuño, R. M.; Guingant, A.; d'Angelo, J. *Tetrahedron Lett.* **1988,** *52*, 6989.

16. Yamamoto, Y.; Furuta, T.; Matsuo, J.; Kurata, T. *J. Org. Chem.* **1991,** *56*, 5737.

17. Jones, W. H.; Mangold, D.; Plieninger, H. *Tetrahedron* **1962,** *18*, 267.

18. Reichardt, C. *Solvent Effects in Organic Chemistry*, Verlag-Chemie, Weinheim, 1979, pp. 144–155.

19. Pearson, R. G.; Sobel, H.; Songstad, J. *J. Am. Chem. Soc.* **1968,** *90*, 319.

20. Smiley, R. A.; Arnold, C. *J. Org. Chem.* **1960,** *25*, 257.

21. Friedman, L.; Schechter, H. *J. Org. Chem.* **1960,** *25*, 877.

22. Fuchs, R.; Mahendran, K. *J. Org. Chem.* **1971,** *36*, 730.

23. Reichardt, C. *Angew. Chem. Internatl. Ed.* **1979,** *18*, 96.

24. Dehmlow, E. V.; Dehmlow, S. S. *Phase Transfer Catalysis*, 2nd ed., Verlag-Chemie, Weinheim, 1983.

25. Makosza, M. "Two Phase Reactions in Organic Chemistry," in *Survey of Progress in Chemistry*, Vol. 9, A. F. Scott, Ed., Academic Press, New York, 1980.

26. Starks, C.; Liotta, C. *Phase Transfer Catalysis, Principles and Techniques*, Academic Press, New York, 1978.

27. Weber, W. P.; Gokel, G. W. *Phase Transfer Catalysis in Organic Synthesis*, Springer Verlag, Berlin, 1977.

28. Orsini, F.; Pelizzoni, F. *J. Org. Chem.* **1980,** *45*, 4726.

29. Crandall, J. K.; Pradat, C. *J. Org. Chem.* **1985,** *50*, 1327.

30. Lee, D. G. *Oxidation in Organic Chemistry*, Vol. 5-D, Trahanovsky, W. S., Ed., Academic Press, New York, 1982, p. 147–206.

31. Lee, D. G.; Lamb, S. E.; Chang, V. S. *Org. Synth.*, **1990,** *Coll. Vol.* VII, 397.

32. Rolla, F. *J. Org. Chem.* **1981,** *46*, 3909.

33. Porter, N. A.; Ziegler, C. B., Jr.; Khouri, F. F.; Roberts, D. H. *J. Org. Chem.* **1985,** *50*, 2252.

34. Raber, D. J.; Guida, W. C. *J. Org. Chem.* **1976,** *41*, 690.

35. Sala, T.; Sargent, M. V. *J. Chem. Soc. Chem. Commun.* **1978,** 253.

36. Landini, D.; Rolla, F. *Synthesis* **1976,** 389.

37. Loupy, A.; Pedoussaut, M.; Sansoulet, J. *J. Org. Chem.* **1986,** *51*, 740.

38. Gokel, G. W.; Durst, H. D. *Synthesis* **1976,** 168–184.

39. Liotta, C. L.; Harris, H. P.; McDermott, M.; Gonzales, T.; Smith, K. *Tetrahedron Lett.* **1974,** 2417.

40. Cook, F. L.; Bowers, C. W.; Liotta, C. L. *J. Org. Chem.* **1974,** *39*, 3416.

41. Rieke, R. D. *Acc. Chem. Res.* **1977,** *10*, 301.

42. Baker, K. V.; Brown, J. M.; Hughes, N.; Skarnulis, A. J.; Sexton, A. *J. Org. Chem.* **1991,** *56*, 698.

43. Salaun, J.; Marguerite, J. *Org. Synth.* **1985,** *63*, 147.

44. Lindley, J.; Mason, T. J. *Chem. Soc. Rev.* **1987,** *16*, 275–311.

45. Suslick, K. S. *Ultrasound, Its Chemical, Physical, and Biological Effects*, VCH Publishers, New York, 1988.

46. Boudjouk, P. *J. Chem. Ed.* **1986,** *63*, 427–429.

47. Uyehara, T.; Yamada, J.; Ogata, K.; Kato, T. *Bull. Chem. Soc. Jpn.* **1985,** *58*, 211.

48. Han, B.-H.; Boudjouk, P. *J. Org. Chem.* **1982,** *47*, 5030.

49. Yamawaki, J.; Sumi, S.; Ando, T.; Hanafusa, T. *Chem. Lett.* **1983,** 379.

50. Regen, S. L.; Singh, A. *J. Org. Chem.* **1982,** *47*, 1587.

51. McKillop, A.; Young, D. W. *Synthesis* **1979,** 401, 481.

52. Posner, G. H. *Angew Chem. Internatl. Ed.* **1978,** *17*, 487.

53. Klein, H.; Steinmetz, A. *Tetrahedron Lett.* **1975,** 4249.

54. Cohen, Z.; Keinan, E.; Mazur, Y.; Varkony, T. H. *J. Org. Chem.* **1975,** *40*, 2141.

55. Cohen, Z.; Varkony, H.; Keinan, E.; Mazur, Y. *Org. Synth.* **1988,** *Coll Vol.* VI, 43.

56. Keinan, E.; Mazur, Y. *J. Org. Chem.* **1977,** *42,* 844.

57. Keinan, E.; Mazur, Y. *J. Am. Chem. Soc.* **1977,** *99,* 3861.

58. Hudlicky, M. *J. Org. Chem.* **1974,** *39,* 3460.

59. Sato, W.; Ikeda, N.; Yamamoto, H. *Chem. Lett.* **1982,** 141.

60. Novak, L.; Poppe, L.; Szantay, C. *Synthesis* **1985,** 939.

61. Nishiguchi, T.; Bougauchi, M. *J. Org. Chem.* **1989,** *54,* 3001.

62. West, P. R.; Davis, G. C. *J. Org. Chem.* **1989,** *54,* 5176.

63. Watson, Jr., H. A.; O'Neill, B. T. *J. Org. Chem.* **1990,** *55,* 2950.

64. Giordano, C.; Cavicchioli, S.; Levi, S.; Villa, M. *J. Org. Chem.* **1991,** *56,* 6114.

65. Feringa, B. L.; de Jong, J. C. *J. Org. Chem.* **1988,** *53,* 1125.

66. Reider, P. J.; Davis, P.; Hughes, D. L.; Grabowski, E. J. J. *J. Org. Chem.* **1987,** *52,* 955.

67. Forbes, C. P.; Wenteler, G. L.; Wiechers, A. *J. Chem. Soc. Perkin I* **1977,** 2353.

68. Eaton, P. E.; Mueller, R. H.; Carlson, G. R.; Cullison, D. A.; Cooper, G. F.; Chou, T. C.; Krebs, E.-P. *J. Am. Chem. Soc.* **1977,** *99,* 2751.

69. Bal, S. A.; Marfat, A.; Helquist, P. *J. Org. Chem.* **1982,** *47,* 5045.

70. Durland, J. R.; Adkins, H. *J. Am. Chem. Soc.* **1939,** *61,* 429.

71. Bennett, J. G.; Bunce, S. C. *J. Org. Chem.* **1960,** *25,* 73.

72. Lee, D. G.; Chang, V. S. *Synthesis* **1978,** 462.

# 10

# INTERPRETATION OF NMR SPECTRA

Most organic compounds are colorless solids or liquids of rather similar appearance, and unknown samples present a puzzle for identification. Physical measurements such as melting point, boiling point, and refractive index are useful for matching against lists of values for limited numbers of known compounds. Actual structural information is readily obtained by means of various spectroscopic methods[1-3] (or ultimately by X-ray crystallography). Ultraviolet–visible spectroscopy gives information on the extent, shape, and substituents of $\pi$-conjugation in molecules. It is a measure of the energy gaps between the electronic ground and excited states. Infrared spectroscopy is particularly useful for determining the presence and identity of functional groups. This is a measure of the frequency of bending and stretching of bonds where the bond dipole changes with the movement. The stretching vibrations of double and triple bonds in alkenes and alkynes involve small bond dipole changes and give weak or no infrared absorptions. For these cases laser Raman spectroscopy gives strong, informative signals.[4]

Mass spectra do not involve electromagnetic radiation as the others do. A molecule is ionized and the resulting ionic fragments are sorted by mass, and their abundance and masses are measured. This is useful for determining what elements are present, and much structural information.[5] It has the special value of requiring as little as one microgram of sample.

All of these methods are important, but here we select to elaborate on the currently most heavily used technique, nuclear magnetic resonance (NMR) spectroscopy because it gives structural information about an entire molecule.[6] It gives the connections that make up the structure by showing neighboring relationships. Elemental nuclei that have an odd mass number and/or an odd atomic number have a magnetic moment, and many of these can be observed in an NMR spectrometer. Those nuclei that show the most practical value are $^1$H, $^{13}$C, $^{19}$F, and $^{31}$P.

## 10.1 ORIGIN OF THE SIGNALS

Modern NMR spectrometers have features that give high resolution spectra with high sensitivity in a matter of seconds. The stages of the process and the behavior of the nuclei are summarized as follows. Before the sample is inserted in the magnetic field of the instrument, the magnetic moments of the nuclei in the sample are oriented randomly (Fig. I*a*). When the sample is inserted in the large magnetic field of the solenoid, the nuclear moments become oriented at a small angle off the axis of the solenoid, slightly more than half of them with the applied field and the rest against the field. The field exerts a force toward alignment with the axis, but since the nuclei are spinning, the result of that force is that they all precess in the same direction around axes parallel with the solenoid axis (Fig. I*b*), a movement like that seen in a tilted gyroscope in a gravitational field. The frequency of the precession (Larmor frequency) is directly proportional to the intensity of the magnetic field. At a field of 7.0462 tesla, protons precess at 300.00 MHz and $^{13}$C nuclei precess at 75.430 MHz. The net magnetic moment (vector sum) of the collection of nuclear moments in the sample may be represented as a unified symbol (Fig. I*b*) along the axis of the applied field. It is nonzero because there is a small excess of nuclei oriented in the lower energy direction, and it is directly along the axis because the nuclei are randomly distributed around their circle of precession (out of phase). Their motion is called *incoherent*.

The axis of the solenoid is called the $z$ axis. The sample is also on the axis of a smaller coil perpendicular to the $z$ axis (Fig. I*c*). An intense pulse of radio frequency energy (300 MHz for a $^1$H spectrum) is applied from an oscillator to the coil, lasting typically 14 $\mu$sec, and of sufficient power to cause 90° of precession around the $x$ axis toward the $x,y$ plane. This moves the net magnetization of the hydrogens from alignment along the $z$ axis to a position in the $x,y$ plane. After the pulse the nuclei are equally distributed in the upper and lower energy states, but the small excess is grouped toward one side of the circle of precession. They all continue to precess around the $z$ axis but now the motion of the small

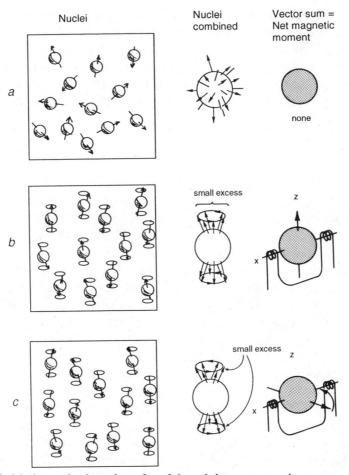

**Figure I.** Motion and orientation of nuclei, and the net magnetic moment. (*a*) before insertion in the instrument solenoid. (*b*) in the solenoid. (*c*) after a 90° pulse. The individual nuclei follow the quantum laws allowing only two states when in a magnetic field (for nuclei of spin quantum number $\frac{1}{2}$), but they can be grouped as parallel sets and give a net magnetic dipole off the z axis. In *a* they are randomly oriented, while in *b* they are oriented all at the same angle off the z axis, randomly around the circle of precession, with a small excess pointing in the upward direction (Boltzman distribution). In *c* the small excess is toward one side, half up and half down, giving a net magnetic dipole in the *x,y* plane bisecting these two groups. The net magnetization is a classical vector and can have any orientation in space.

excess is coherent. Thus the net magnetization now goes around the z axis at 300 MHz inducing a 300-MHz alternating current in the coil on the x axis, which is now connected to the receiver. This current dies out over a few seconds as the nuclei gradually fan out around the z axis and the net magnetization also equilibrates back out of the *x,y* plane toward the z axis, a process called *relaxation*. With relaxation, the excess in the lower energy state is restored, and the precessional motion becomes incoherent.

Why does the oscillating field from the transmitter coil cause precession? The oscillating field is equivalent to a pair of fields rotating around the z axis in the x,y plane in opposite directions, alternately reinforcing and canceling each other at 300 MHz. The one going in the same direction as the precessing nuclei keeps up with those nuclei applying a constant field perpendicular to the z axis, causing the net magnetization to precess away from the z axis. A conventional way to picture this is to imagine yourself on a tiny turntable in the x,y plane going around the net magnetization at 300 MHz. Looking toward the center, the precession would not be evident, just as the rotation of the earth is not obvious to us going around on it, and there is no evidence for the large magnetic field along the z axis. During the pulse you would observe a constant magnetic field directed from you toward the center in the x,y plane. You would then see the net magnetization precess away from the z axis toward your right until it reached the x,y plane when the pulse is turned off.

From the foregoing description, you might expect a one-line spectrum; however, the net magnetic field strength at each hydrogen nucleus in various positions in the molecule is slightly more or less than that provided by the solenoid. This is the result of deshielding or shielding by electrons and neighboring magnetic nuclei. Thus in the combined symbol representation they precess in groups at slightly more or less than the exact frequency of the 300-MHz oscillator, inducing as many frequencies simultaneously in the receiver coil. Figure Ic for a multipeak spectrum would show several net magnetic moments precessing at different rotational rates. A plot of induced potential versus time for this raw result, called the *free induction decay* (FID), would be a complex sum of sine waves (interference pattern) tapering to nothing over a period of seconds. A simple example of a FID is shown in Fig. II. For the actual recorded FID, the oscillator frequency is fed to the receiver and the differences between the oscillator and the precessing moments is recorded; thus, it is a pattern of lower ($0$–$10^4$ Hz) frequencies with substantial rather than miniscule differences.

A spectrum is a plot of intensity versus frequency. The conversion of the FID to a spectrum is done by a mathematical process known as a *Fourier transform*, carried out swiftly by the instrument computer. The spectrum is conventionally plotted with precessional frequency decreasing toward the right along the x axis. The scale is in units of parts per million (ppm) difference from a standard, usually tetramethylsilane

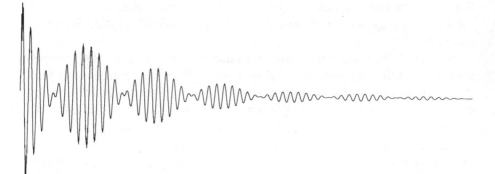

**Figure II.** A FID plot for 2-propanol. The strong doublet from the $(CH_3)_2$ consists of two frequencies that, because they are equally intense, completely cancel each other in those moments when they are 180° out of phase; thus, the plot appears as a rising and falling series of waves that periodically reduce to zero. They reinforce and destructively interfere about every nine cycles with this particular oscillator offset. The weaker frequencies for the CHOH can be seen superimposed on this major pattern.

(TMS) with the signals to the left of TMS (less shielded nuclei) given positive numbers, and those to the right of TMS (more shielded), given negative. The frequency differences from TMS were converted to ppm by multiplying by 1 ppm/300 Hz in the case of a 300 MHz spectrum. A well-tuned spectrometer can resolve proton signals differing by less than 0.3 Hz out of the 300 MHz.

If the pulse from the oscillator were one pure frequency, it would interact with only those nuclei that precessed at that one frequency rather than all the protons in different environments. Actually, a very short pulse with a sharp onset and cessation provides a range of frequencies equal to $\pm$ the reciprocal of the duration of the pulse; for example, 12 $\mu$sec gives $\pm 83$ kHz, far more than enough to cover all chemical-shift possibilities.

Molecular transitions in the ultraviolet range (electronic excitation) are of such high energy that nearly all the molecules are in the lower-energy state at room temperature, and are ready to absorb radiation; therefore, the spectroscopic signals are strong. The transitions in NMR are of such low energy that at room-temperature thermal equilibrium, the population excess in the lower-energy states is miniscule (3 in $10^5$ at 7 tesla). This leads to far weaker signals requiring great amplification, which produces a considerable amount of random background noise. This is overcome by repeating the pulse–relaxation sequence many times to accumulate data and improve the signal/noise ratio. This is useful for obtaining clear spectra from very small samples. It is a necessity for obtaining $^{13}C$ spectra since the signals are intrinsically very weak and the natural abundance is low. Doubling the number of observed FIDs increases the signal to noise ratio by a factor of 1.414.

This is a highly simplified description of the whole process. For more thorough descriptions, see the refs.[7-10]

## 10.2 INTERPRETATION OF PROTON NMR SPECTRA

Figure III illustrates the 300 MHz $^1$H spectrum of suberanilic acid. All the signals may be assigned to particular hydrogens in the structure using the chemical-shift values, the splitting multiplicity, and comparison with spectra of similar compounds:

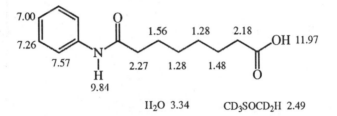

In the introductory course you learned to use four basic kinds of information from $^1$H NMR spectra: number of signals, chemical-shift values, integrated signal area, and splitting patterns. We will now delve further into some of these, particularly the splitting patterns.

### 10.2.1 Coupling Constants and Connectivity

When two signals are coupled with each other, there will be an identical coupling constant as part of the multiplicity of each signal. Coupling constants can often be measured to two or three significant figures (to about 0.1 Hz), which allows matching values in various multiplets, in order to discover coupled neighbors. An equivalent set of $n$ neighbors will provide $n + 1$ subpeaks in a signal, all equally spaced, and with area distribution following the binomial distribution. For example, three equivalent neighbors will split a signal into four peaks of relative area $1:3:3:1$, called a *quartet*. Nonequivalent neighbors will split with different coupling constants and give up to $2^n$ subpeaks. For example, three all-different neighbors will give a doublet of double doublets (ddd), which will appear as eight peaks all with the same areas. This may be diagramed as an inverted tree as in Fig. IV. The ddd will include three different coupling constants, which can be measured as follows. On command the instrument will give exact ppm values to the maxima in

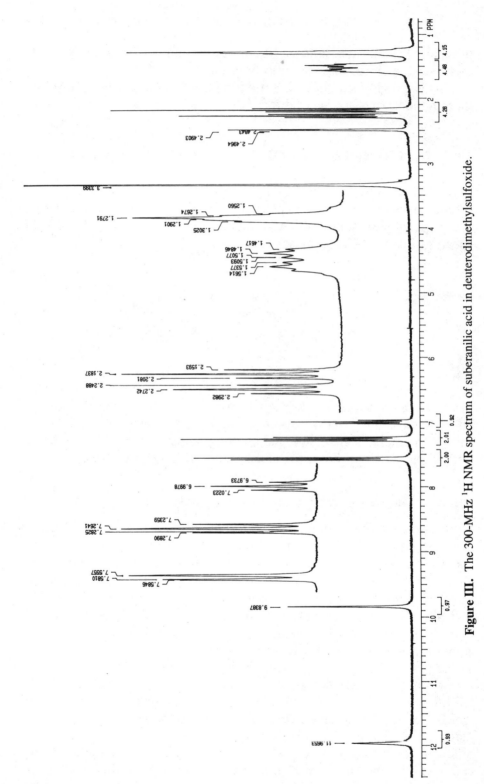

**Figure III.** The 300-MHz $^1$H NMR spectrum of suberanilic acid in deuterodimethylsulfoxide.

294

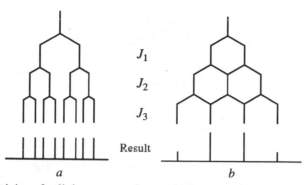

**Figure IV.** Origins of splitting patterns in a multiplet. In *a* the three coupling constants are different. Starting with a single line for the unsplit signal, we first apply $J_1$, which gives a doublet; $J_2$ then doubles the doublet, and $J_3$ doubles that, resulting in a total of eight lines, all of equal height. In *b* the three coupling constants are equal (because the three neighboring hydrogens are equivalently arranged) and there are coincidences of lines that increase the probability of the resulting two inner lines.

the multiplet. The smallest $J$ value is obtained by (counting from the left) subtracting the value of the second line from the first (or fourth from third, etc.) and multiplying by Hz/ppm (300 if it is a 300-MHz spectrometer). The next larger $J$ value is obtained by subtracting the third from first (or fourth from second, etc.), and the largest $J$ by subtracting the fifth from first (or sixth from second etc.). The $J$ value in the quartet is obtained by simply subtracting any two adjacent lines and multiplying by Hz/ppm.

Any combination of neighboring protons might exist causing splitting patterns such as dd, dt, dq, tq, and so on. Each of these may be diagramed with an inverted tree, and the $J$ values obtained from appropriate subtractions. This is illustrated with the spectrum of *N*-allyl-*o*-toluidine (Fig. V), where the $J$ values are obtained, and matched in order to pair the neighbors and assign the multiplets to particular hydrogens in the structure. For example, the multiplet at 3.8 ppm is a doublet of triplets. The doublet coupling constant is found by taking the difference between the middle line of each triplet and multiplying by 300 Hz/ppm: (3.8294 − 3.8116) × 300 = 5.34. The triplet coupling constant is found by taking the difference between any adjacent pair of lines within one of the triplets: (3.8346 − 3.8294) × 300 = 1.56. The results of similar analyses of all the multiplets are given in Table I. By matching coupling constants we find that 3.82 is coupled with 5.18, 5.29, and 6.00. The triplet is a coincidence of coupling constant with 5.18 and 5.29. Likewise, we find that those first four multiplets are all coupled together. Multiplet 6.66 is coupled with 7.06, and a proton meta to it, but the

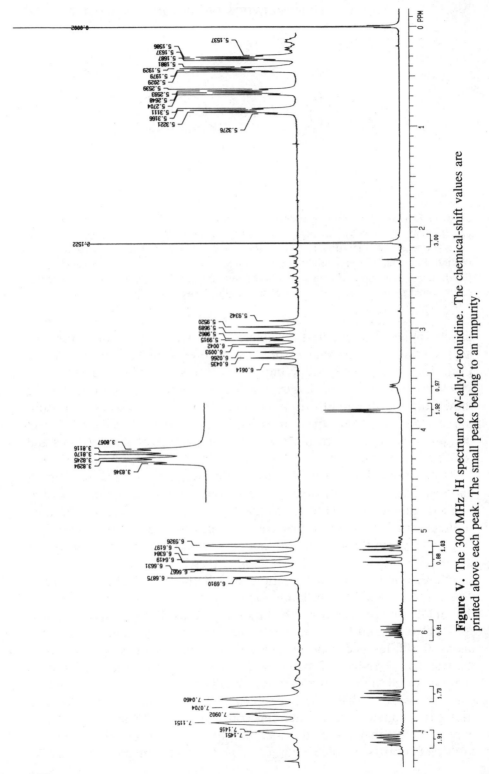

**Figure V.** The 300 MHz ¹H spectrum of *N*-allyl-*o*-toluidine. The chemical-shift values are printed above each peak. The small peaks belong to an impurity.

296

**TABLE I. Multiplicities in the ¹H NMR Spectrum of N-Allyl-o-toluidine**

| Chemical Shift | Multiplicity | J values | | |
|---|---|---|---|---|
| 3.82 | dt | 5.3 | 1.6 | |
| 5.18 | dq | 10.3 | 1.5 | |
| 5.29 | dq | 17.2 | 1.6 | |
| 6.00 | ddt | 10.3 | 17.2 | 5.3 |
| 6.61 | d | 8.1 | | |
| 6.66 | dt | 1.0 | 7.3 | |
| 7.06 | d | 7.3 | | |
| 7.12 | dt | 1.0 | 8.0 | |

small coupling constant is not resolved in the meta partner. The others are paired accordingly. Assignments of signals to particular hydrogens in the structure may now be made, using these coupling relationships, integral values, and chemical shifts. For example, the larger coupling constant at 5.29 indicates that it is cis to the $CH_2$ group. Similar analyses of spectra of unknown compounds often allow one to draw a structure with a bonding scheme connecting all the atoms in coupled sets, taking the structural significance of the size of the coupling constants into account.

## 10.2.2  The Magnitude of Coupling Constants

The size of the magnetic influence a proton receives from a neighboring proton depends on the distance between them, the intervening bonds and their angular relationship. The coupling constant $J$ is a measure of this effect. It is specified in units of hertz, which do not vary with different applied field strengths. Typical values of coupling constants for various neighboring relationships are given in Table II.

## 10.2.3  Non-First-Order Signals

When we see a simple triplet, we conclude that the hydrogen(s) giving that signal have two equivalent neighboring hydrogens. When we see a dd, we conclude that the hydrogen(s) giving that signal has (have) two nonequivalent neighbors. This is called *first-order analysis* and is possible when the chemical-shift difference between neighboring hydrogens is much larger than the $J$ value. Often this is not the case. If the chemical-shift difference is less than 10 times the $J$ value, the signals are distorted from simple first-order expectations. This is illustrated with computer simulations in Figs. VI, VII, and VIII.

**TABLE II. Coupling Constants for Neighboring Hydrogens (Hz)**

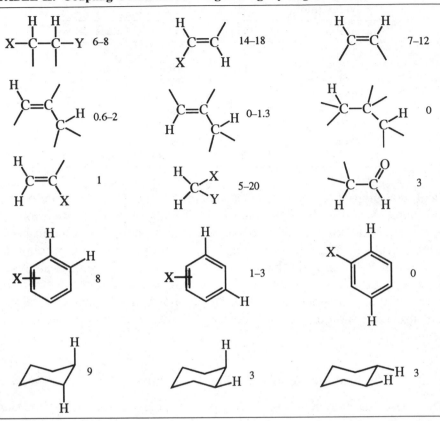

In Fig. VI we have a single hydrogen with its single neighbor (AB). In the bottom spectrum we recognize a pair of doublets. As the chemical-shift difference becomes smaller, proceeding to the top of Fig. VI, the inner half of each doublet becomes taller while the outer diminishes. The extreme would be identical neighboring hydrogens, which would, of course, be one singlet, which these spectra approach. One or both of these signals may be resplit by other neighbors, but the distorted doublet aspect of it should be discernable, with a corresponding $J$ value, even if it is a dd, dt, or dq. No attempt is made here to give the theoretical reasoning or predictive methods for these non-first-order patterns,[11, 12] but you should be able to *recognize* them as you analyze spectra, and know what structural meaning they have.

In Fig. VII at the bottom we see a first-order spectrum for a $CH_2$ group and a single neighbor ($A_2B$). In the progression toward smaller chemical-shift differences we see distortions of size and increasing num-

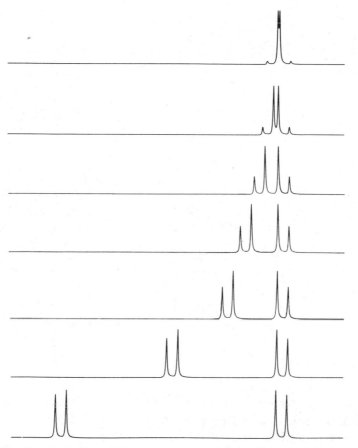

**Figure VI.** Simulated AB splitting pattern with $J/\Delta\nu$ = 1.5, 1, 0.5, 0.3, 0.2, 0.1, and 0.05. Courtesy of R. F. Evilia, University of New Orleans.

bers of peaks, again trending at the top toward a singlet. In Fig. VIII we see a similar series for two neighboring $CH_2$ groups ($A_2B_2$). Notice that the left and right halves of the signals are mirror images, which makes this pattern quickly recognizable in spectra. More complicated patterns involving larger numbers of hydrogens have been analyzed and computer-simulated.

## 10.2.4   Spin Decoupling

In some spectra it may not be possible to distinguish coupling constants, or several $J$ values may be essentially the same, and conclusions about neighboring relationships may not be possible. If a second oscillator is

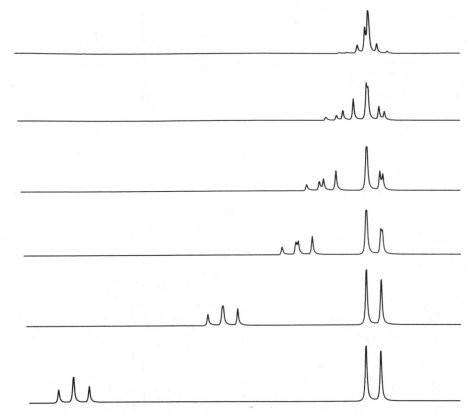

**Figure VII.** Simulated $A_2B$ splitting pattern with $J/\Delta\nu = 1, 0.5, 0.3, 0.2, 0.1$, and 0.05. Courtesy of R. F. Evilia, University of New Orleans.

placed to continually irradiate at one particular signal, it will cause rapid spin inversions that will average the magnetic influence of those nuclei to zero. If at the same time the rest of the spectrum is obtained, the splitting of neighbors normally caused by the nuclei that are under continuous irradiation will disappear, thus identifying those neighbors. This is sometimes referred to as *double resonance*.

### 10.2.5 The Nuclear Overhauser Effect (NOE)

Other neighboring relationships may be traced by relaxation transfer. The spin–lattice relaxation of excited nuclei toward equilibrium is the transfer of polarization to other nearby nuclei. If an oscillator is set at one spectral signal to disturb the equilibrium populations, the relaxation of these populations involves simultaneous inversions of other nuclei nearby. This results in more nuclei in the lower energy state and thus

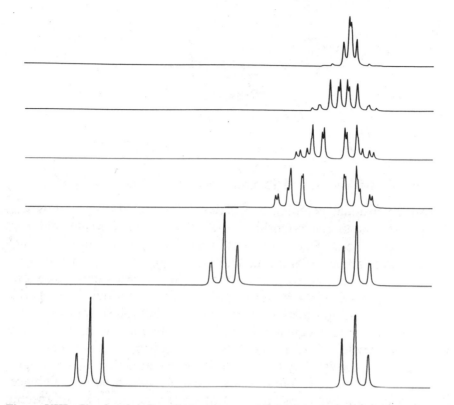

**Figure VIII.** Simulated $A_2B_2$ splitting pattern with $J/\Delta\nu = 1, 0.5, 0.3, 0.2, 0.1,$ and 0.05. Courtesy of R. F. Evilia, University of New Orleans.

stronger signals from those neighbors during the irradiation of the first signal. This enhancement is a through-space effect determined by the distance to the neighbor, and not by intervening bonds. The effect falls off steeply with the distance; it is proportional to $1/r^6$. This effect is called the *nuclear Overhauser effect*. In practice, a spectrum is run, and then another is run during irradiation of a selected signal. The normal spectrum is subtracted from the latter to reduce all signals to zero except those that were enhanced by the irradiation. This is called a *NOE difference spectrum*. Those enhanced signals indicate proximity to the hydrogen whose signal was irradiated.

The stereochemistry at the carbon carrying the OH group in compound **1** was determined with a NOE difference spectrum.[13] Irradiation at the signal for the hydrogen on that carbon gave enhancement to a hydrogen on the [1] bridge and the nearest hydrogen on the aromatic ring. Thus the OH must be nearer the [2] bridge and the compound is the endo isomer.

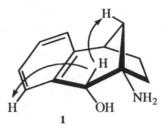

1

## 10.2.6 Two-Dimensional Proton Correlation Spectroscopy

A two-dimensional coupling correlation [1]H NMR spectrum shows all the proton coupling relationships in a molecule in a single, although longer, experiment. This is more efficient than spin decoupling, which probes one relationship at a time and may require many experiments.

Two-dimensional correlations are brought out with multipulse irradiations.[14] For a COSY experiment, a 90° pulse is followed by a period of precession $(t_1)$ and then a second (mixing) 90° pulse transfers magnetization including frequencies between neighbors among the precessing sets of nuclei. This is followed by acquisition of the FID. The sequence is repeated many times with incremental increases in $t_1$ to give a stack of spectra after Fourier transformation. Precessional differences during $t_1$ cause a cyclic rise and fall of signals in the series of spectra. A set of Fourier transformations through the stack of spectra at increments in the spectral frequency $F1$ produces a second frequency scale $F2$. A two-dimensional plot of $F1$ against $F2$ gives signals that are represented by contour lines indicating intensity of the signal. On these plots there are signals that have both frequencies of those pairs of protons that are coupled. Although this process is lengthy and complicated, COSY spectra are readily available from automated spectrometers, and the simplicity of interpretation makes them very useful. How these plots are obtained will not be covered here, but interpretation is presented.

In the COSY spectrum of 3-heptanone (Fig. IX) all the neighboring relationships may be traced by locating the two chemical-shift values correlated by the off-diagonal signals. The diagonal in this figure runs from the upper left to the lower right. The five signals on the diagonal are not useful because they correlate a signal in $F1$ with itself in $F2$; however, those off the diagonal (duplicated on each side of the diagonal) give the neighboring relationships. Start with the off-diagonal signal at $F1 = 0.7$ and $F2 = 1.15$. The highest field triplet is the methyl group on carbon 7, and that off-diagonal signal indicates that the multiplet at 1.15 is the neighboring $CH_2$ group on carbon 6. The multiplet at 1.15

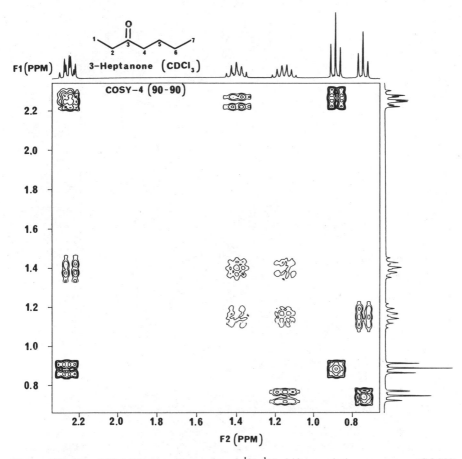

**Figure IX.** The 300-MHz two-dimensional ¹H–¹H shift correlation spectrum (COSY) of 3-heptanone. Spectrum courtesy of Varian Associates.

has another off-diagonal signal at 1.4, indicating that the signal at 1.4 is the neighbor on the other side, the CH$_2$ on carbon 5. That signal at 1.4 has another off-diagonal signal at 2.22, which must be the neighbor on carbon 4. Thus moving along from one off-diagonal signal to another establishes the connectivity in the structure. Of course, the assignments in 3-heptanone could have been made with only a one-dimensional spectrum using chemical-shift considerations, but this simple example was chosen for illustration. They are more useful for complex molecules as illustrated in many recent journal articles.

Figure X shows a COSY spectrum of 2-vinylpyridine. The highest field signals are for the hydrogens of the CH$_2$ group. Starting here, you can trace all the neighboring relationships and assign all the signals. The

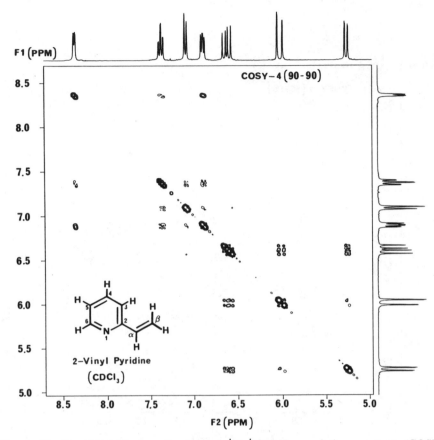

**Figure X.** The 300-MHz two-dimensional $^1$H–$^1$H shift correlation spectrum (COSY) of 2-vinylpyridine. Spectrum courtesy of Varian Associates.

multiplicity of the signals will confirm some of your assignments. The $\alpha$ hydrogen is coupled with the hydrogen on position 3 of the ring, but as you might expect, the coupling constant is very small. The off-diagonal signal is thus very small, but visible and certainly usable. Small signals generally indicate small coupling constants and/or coupling with multiplets that do not have much height.

Long-range COSY spectra may be obtained by allowing a delay for further precession before the mixing pulse and before the acquisition each time. This allows coupled pairs correlated by small values of $J$ to give appreciable off-diagonal signals. In this way connectivity can be traced over three or four bonds. For good examples, see the work of Cho and Harvey.[15]

For complex molecules that give unresolved $^1$H spectra, cross sections of COSY spectra may be taken, giving a set of one-dimensional spectra.

In these, signals that overlap in the conventional $^1H$ spectrum are often separated in different cross sections where they can be seen clearly for their multiplet structure. A striking example is given in the work of Portlock et al.[16]

A two-dimensional relaxation spectrum (nuclear Overhauser effect spectroscopy, NOESY) shows all the NOE relationships in a molecule in one multipulse experiment. The off-diagonal signals relate those hydrogens that are close through space. NOESY spectra are difficult to produce because they require careful optimization of oscillator power and frequency; however, they have been used to answer many structural questions. For an example that allowed many assignments within a complex structure, see Cushman et al.[17]

The many other kinds of two-dimensional spectra may be found in the general references.

### 10.2.7  Spectra at Higher Magnetic Fields

Equation 1 shows the direct proportionality between the magnetic field strength and the oscillator frequency for proton resonance, where $\nu$ is the frequency in hertz and $B_0$ is the magnetic field strength in tesla.

$$\nu = 4.2576 \times 10^7 \, B_0 \qquad (1)$$

In the early days $^1H$ NMR spectra were usually run at 60 MHz in a magnetic field of 1.4092 tesla, while now 300 MHz or higher is common. What is gained at higher magnetic fields? The magnetic field of the instrument orients the motion of electrons in molecules such that they give local opposing magnetic fields. Protons near electron-donating groups have larger induced fields shielding them and appear farther to the right in a spectrum while protons with electron-withdrawing groups nearby have smaller induced fields around them and appear farther to the left in a spectrum. Doubling the instrument magnetic field doubles the difference between the induced fields at each nucleus; thus the frequency difference between the signals will be doubled, and the spectrum is more spread out. If we simply stretch a 60-MHz spectrum out horizontally, we will not see more peaks or eliminate overlaps; however, in the high-field spectra the overlaps are lessened. The width of a multiplet is determined by the magnetic field effects of nearby nuclei, the size of which is fixed and independent of the instrument magnetic field. Therefore, as the spectrum is spread with higher magnetic fields, the individual multiplets do not widen but come away from each other, diminishing their overlaps, and facilitating interpretation.

The chemical-shift axis of spectra is not generally in units of frequency but in parts per million. That is ($\Delta\nu$ between signals) $\times 10^6 \div$ (oscillator frequency). Thus when we use a magnet of twice the field strength, the $\Delta\nu$ between signals doubles but the dividend remains the same. This is done so that chemical-shift differences have the same ppm values regardless of the applied magnetic field. On the other hand, the $J$ values remain the same in Hz or tesla as we go from one instrument to another. Therefore, on the ppm scale the separation of peaks within a multiplet shrinks to half as we go to an instrument with twice the magnetic field strength. Thus it appears that with increasing magnetic field, the multiplets are narrowing but staying centered on the same ppm values (Fig. XI). Spectacular detail has been resolved in spectra of complex molecules at high fields.

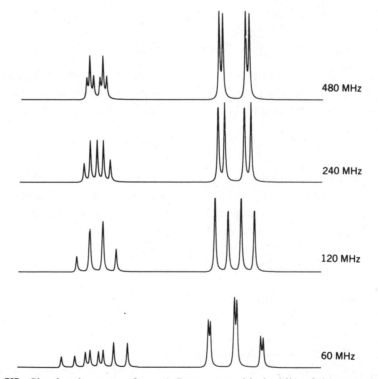

**Figure XI.** Simulated spectra of two $A_2B$ patterns with doubling field strengths. This is an arbitrary selection of two patterns that overlap with coincidences at lower field strengths. The lowest trace is the appearance at 60 MHz where peaks are coinciding in the $A_2$ portion. The distortion from first order is equivalent to that in the $J/\Delta\nu = 0.2$ trace in Fig. VII. The B portion is clearly two intermingled patterns, each the same as that in Fig. VII. The 120-MHz trace is first order interpretable, but the two triplets in the B portion are now doubly coinciding. The 240-MHz trace lessens the overlaps, and finally the 480-MHz spectrum (top) is clear with no overlaps. All four spectra are plotted to the same ppm scale. These simulations were provided by R. F. Evilia, University of New Orleans.

There is another major change in spectra on going to higher-field instruments. The ratio of chemical shift to $J$ increases for all coupled pairs; therefore, the signals become more nearly first-order and interpretation is simplified.

Finally, the higher magnetic field increases the energy difference between the upper- and lower-energy states of the nuclei, and the Boltzman distribution will give a larger excess of nuclei in the lower state at equilibrium. This provides a substantial increase in the strength of the signals.

## 10.2.8  Stereochemical Effects

Diastereomers are chemically different materials with different physical properties, and they give different spectra. The differences may be small but are often sufficient for measuring the percentage of each isomer in mixtures by integration.

Diastereotopic groups reside in diastereomeric environments in molecules (Section 3.8) and thus give separate signals, and each can split the signal of the other if they are close together. In the spectrum of acrolein diethyl acetal (Fig. XII), notice that the $CH_2$ groups do not give a simple quartet as they do in acetone diethylacetal. Each hydrogen of a $CH_2$ group gives a separate signal. Each signal is split by the diastereotopic hydrogen on the same carbon (with distortion as in Fig. VI) and also by the three neighboring hydrogens on the next carbon. The result is two doublets of quartets, 16 lines in all. The other equivalent $CH_2$ group coincides, giving two signals of area 2H each. When a signal appears more complex than it ought to on simple considerations and is also nearly symmetric, consider diastereotopic groups. If they are geminal hydrogens, they should split each other with considerable distortion, and you should look for the small outer peaks that complete the pattern.

Enantiomeric compounds give identical spectra in ordinary solvents, but diastereomeric complexes may form in the presence of a single enantiomer of a chiral complexing agent. Therefore, enantiomers may give separate measurable signals as discussed in Section 3.3. Enantiotopic groups or atoms in a molecule will also give separate signals in chiral media. For example, a chiral praseodimium reagent (Chapter 3, structure **13** with Pr in place of Eu) added to a solution of benzyl alcohol in $CCl_4$ gave two doublets, one for each enantiotopic hydrogen ($J = 13.0$ Hz, $\Delta\nu = 0.13$ ppm).[18] Since it is a shift reagent, it also moved the signals 6.63 ppm to the right. It gave two singlets for the enantiotopic methyl groups in dimethylsulfoxide, and also distinguished the methyls in 2-propanol.

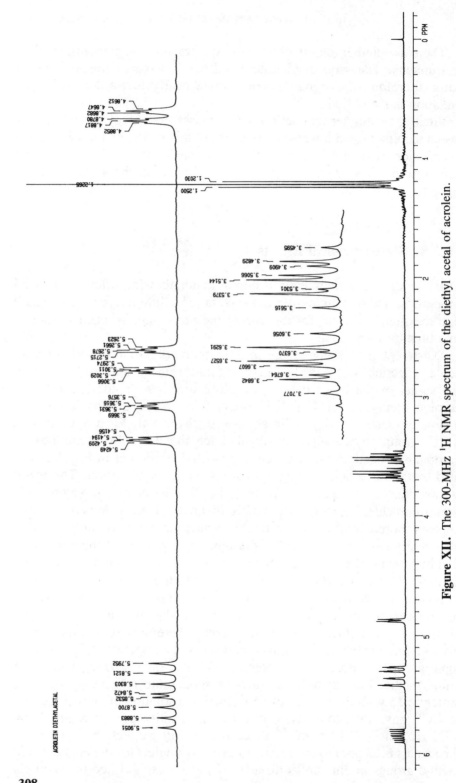

**Figure XII.** The 300-MHz $^1$H NMR spectrum of the diethyl acetal of acrolein.

ACROLEIN DIETHYLACETAL

4.8612
4.8647
4.8682
4.8780
4.8817
4.8852

1.2566
1.2030
1.2500

3.4595
3.4829
3.4909
3.5066
3.5144
3.5301
3.5379
3.5516
3.6056
3.6291
3.6370
3.6527
3.6607
3.6764
3.6842
3.7077

5.2692
5.2661
5.2678
5.2715
5.2794
5.3014
5.3029
5.3066
5.3576
5.3616
5.3631
5.3669
5.4156
5.4194
5.4209
5.4249

5.7952
5.8121
5.8303
5.8472
5.8532
5.8700
5.8883
5.9051

## 10.3  INTERPRETATION OF CARBON NMR SPECTRA

The abundant $^{12}$C isotope has no nuclear magnetic moment, but the 1.1% natural abundance $^{13}$C isotope does.[19] The signal from these carbons, however, is only $\frac{1}{64}$th the intensity from the same number of hydrogens. These two factors cause a carbon spectrum to be about $\frac{1}{6000}$th the intensity of a hydrogen spectrum of the same sample. This necessitates the use of pulsed irradiation and Fourier transform methods and summation of many repeated spectra (Section 10.1). Typically the spectra are obtained at 2–10-sec intervals and, for a routine 5–20-mg sample, less than 15 min is required to accumulate a good summation.

### 10.3.1  General Characteristics

*Chemical-Shift Range* The signals for carbon occur $2.98 \times 10^6$ ppm to the right of those from protons. At 7.0462 tesla the frequency for carbon resonance is 75.430 MHz, and for hydrogen it is 300.00 MHz. The chemical-shift values are generally recorded as ppm from the carbons of tetramethylsilane and extend over a range of about 230 ppm. Since this is more than 10 times the range found for hydrogens, the problem of obscuring overlaps is much less here. The $sp^3$ hybrid carbons bonded only to hydrogens and other carbons generally occur in the 6–50-ppm range. Those carrying electronegative atoms give signals 10–60 ppm greater. The $sp^2$ hybrid carbons of alkenes and aromatic rings occur in the 100–165-ppm range, and those in carbonyl groups are at 155–220 ppm. The $sp$ hybrid carbons in alkynes are at 65–85 ppm. In Section 10.3.2 we will see how to make good numerical predictions for specific structures.

*Number of Signals* Each structurally different carbon in a molecule gives a separate signal. Identical carbons will, of course, coincide. A *tert*-butyl group gives two signals, one for the three equivalent $CH_3$ carbons and one for the central carbon. *p*-Nitroanisole gives five signals, two of which represent two carbons each.

*Splitting of Signals* The low natural abundance of $^{13}$C makes it highly improbable that two would be side by side in a molecule; therefore, $^{13}$C–$^{13}$C splitting is not observed in routine spectra. Protons split the signals of the carbon to which they are bonded by 100–240 Hz. Protons on an adjacent or the next farther carbon give only 4–6-Hz splitting. The lower spectrum in Fig. XIII illustrates proton splitting. The quartet centered at 18.9 ppm indicates three directly bonded hydrogens, that is,

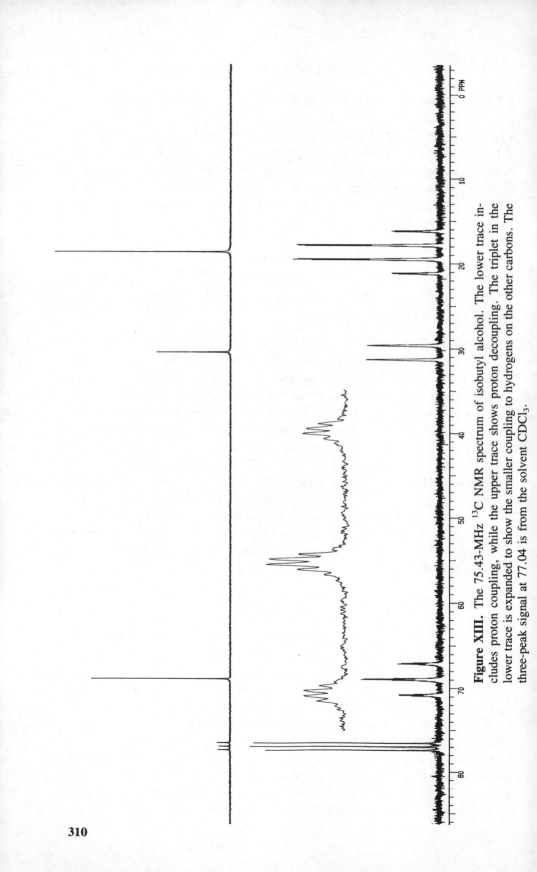

**Figure XIII.** The 75.43-MHz $^{13}C$ NMR spectrum of isobutyl alcohol. The lower trace includes proton coupling, while the upper trace shows proton decoupling. The triplet in the lower trace is expanded to show the smaller coupling to hydrogens on the other carbons. The three-peak signal at 77.04 is from the solvent $CDCl_3$.

a methyl group. The finer splitting is caused by neighboring hydrogens on $\alpha$ and $\beta$ carbons. The doublet at 30.8 indicates a CH group, and finally, the triplet at 69.4 is a $CH_2$ group. Without resolving the very small splitting, one may simply generalize that quartets are $CH_3$ groups, triplets are $CH_2$ groups, doublets are CH groups, and singlets are carbons bearing no hydrogens. This is very simple compared to proton spectra, but the large coupling constants make it difficult to know which peaks belong together as multiplets in complex spectra. Routinely, broad irradiation (spin decoupling) of all the protons is used to reduce the $^{13}C$ spectrum to all singlets, as in the upper trace in Fig. XIII. This provides an advantage in increased sensitivity as well, because a signal that is concentrated entirely in a sharp singlet stands taller, giving a far greater signal/noise ratio.

*Area of Signals* During the accumulation of spectra at short intervals, the excited $^{13}C$ nuclei relax toward equilibrium distribution after each pulse. Each carbon in a structure does this at a different rate, and those with no hydrogen atoms bonded to them are particularly slow. Those less completely relaxed give less signal in the succeeding pulse, which leads to accumulated signals of varying intensities where those from carbons bearing no hydrogen are weak. Thus the signals areas are not quite proportional to the number of carbons they represent in the structure. If desired, long intervals between pulses may be used to obtain signal areas nearly proportional to the ratio of carbons represented, but this may require excessive instrument use time.

The broad irradiation of the proton resonances for spin decoupling also causes relaxation transfer (NOE) to the carbons to which they are attached. This causes an increase in the proportion of $^{13}C$ nuclei in the lower-energy state beyond that in thermal equilibrium, leading to a beneficial signal enhancement up to threefold. This varies according to the nearness of hydrogens, and those carbons bearing no hydrogens give relatively weak signals.

Both the variation in the relaxation time and the NOE result in signal areas that are not proportional to the number of atoms represented; therefore, integrals from $^{13}C$ spectra are not often used. Nevertheless, qualitatively a larger signal may represent twice as many carbons as a smaller signal, and very small signals are often assignable to carbons that bear no hydrogens.

In Fig. XIV we see a routine $^1H$-decoupled $^{13}C$ NMR spectrum of 2-(3-bromopropyl)-4,4,5,5-tetramethyl-1,3-dioxolane. The signals representing pairs of equivalent methyl carbons are larger than those representing one carbon each.

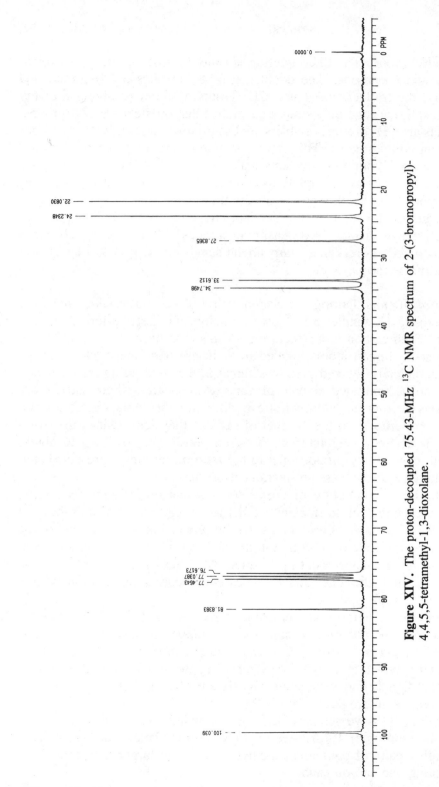

**Figure XIV.** The proton-decoupled 75.43-MHz $^{13}$C NMR spectrum of 2-(3-bromopropyl)-4,4,5,5-tetramethyl-1,3-dioxolane.

The instruments maintain a constant calibration by using a deuterium signal to lock the field–frequency ratio. Therefore, deuterated solvents are necessary. Deuterochloroform is most frequently used. The carbon in this solvent gives a weak signal that is split by the deuterium (which has three spin states) into three equal peaks at 76.62, 77.04, and 77.46 ppm from tetramethylsilane (Fig. XIV). These are seen in many spectra, with greater relative intensity where the solution is dilute. Deuterodimethylsulfoxide is also commonly used, taking care to keep it dry. The small amount of the pentadeutero compound present gives a septuplet at 39.7 ppm.

The individual signals in a proton-spin-decoupled $^{13}C$ spectrum may be identified as arising from a CH, $CH_2$, or $CH_3$ carbon with a distortionless enhancement of polarization transfer (DEPT) experiment. A DEPT experiment is a five-pulse sequence that generates individual $^{13}C$ CH, $CH_2$, and $CH_3$ subspectra.[20]

## 10.3.2  Calculation of $^{13}C$ Chemical-Shift Values

The chemical-shift value for a carbon increases by about 8 ppm for each carbon bonded to it or to an adjacent carbon. An electronegative atom substantially changes the chemical shift of the carbon to which it is attached but has a relatively small effect on nearby carbons. Predictions of the chemical-shift values can be made by using tabulated additive corrections and are useful for assigning signals to particular carbons in a structure. They can help to determine which of several structures is in accord with the spectrum. The corrections are empirically derived from measured values of many known compounds. The chemical-shift values for each carbon in acyclic alkanes and alkyl parts of functional molecules may be calculated by use of the increments in Table III.[21] The error is usually less than 1 ppm.

To illustrate the use of Table III, the values expected from 3-methylpentane are calculated as follows. The C-1 is a $CH_3$ group, and there is one $\alpha$ $CH_2$, one $\beta$ carbon, two $\gamma$ carbons and one $\delta$ carbon.

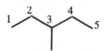

| Increments | Calculated | Measured |
|---|---|---|
| C-1:    $6.8 + 9.6 + 2(-3.0) + 0.5$ | $= 10.9$ | 11.4 |
| C-2:   $15.3 + 16.7 - 2.7$ | $= 29.3$ | 29.3 |
| C-3:   $23.5 + 2(6.6)$ | $= 36.7$ | 36.4 |
| 3-$CH_3$: $6.8 + 17.8 + 2(-3.0)$ | $= 18.6$ | 18.8 |

**TABLE III. Structural Parameters for Calculating $^{13}$C NMR Shifts of Alkanes**

$$\overset{\alpha\quad\beta\quad\gamma\quad\delta}{\text{©—C—C—C—C}}$$

| Carbon to Be Predicted | Add | For Each Carbon of the Indicated Type, Add the Listed Increment | | | | | | | |
|---|---|---|---|---|---|---|---|---|---|
| CH$_3$ | 6.8 | $\alpha$ | CH$_3$ | 0 | $\beta$ | 0 | $\gamma$ | −3.0 | $\delta$ 0.5 |
| | | $\alpha$ | CH$_2$ | 9.6 | | | | | |
| | | $\alpha$ | CH | 17.8 | | | | | |
| | | $\alpha$ | C | 25.5 | | | | | |
| CH$_2$ | 15.3 | $\alpha$ | CH$_3$ | 0 | $\beta$ | 0 | $\gamma$ | −2.7 | $\delta$ 0.3 |
| | | $\alpha$ | CH$_2$ | 9.8 | | | | | |
| | | $\alpha$ | CH | 16.7 | | | | | |
| | | $\alpha$ | C | 21.4 | | | | | |
| CH | 23.5 | $\alpha$ | CH$_3$ | 0 | $\beta$ | 0 | $\gamma$ | −2.1 | $\delta$ 0 |
| | | $\alpha$ | CH$_2$ | 6.6 | | | | | |
| | | $\alpha$ | CH | 11.1 | | | | | |
| | | $\alpha$ | C | 14.7 | | | | | |
| C | 27.8 | $\alpha$ | CH$_3$ | 0 | $\beta$ | 0 | $\gamma$ | −0.9 | $\delta$ 0 |
| | | $\alpha$ | CH$_2$ | 2.3 | | | | | |
| | | $\alpha$ | CH | 4.0 | | | | | |
| | | $\alpha$ | C | 7.4 | | | | | |

2,3-Dimethylbutane is calculated similarly:

| | | | |
|---|---|---|---|
| C-1: | 6.8 + 17.8 + 2(−3.0) | = 18.6 | 19.5 |
| C-2: | 23.5 + 11.1 | = 34.6 | 34.0 |

The vinyl carbons of alkenes are calculated according to Table IV. The error is usually less than 3 ppm. For each carbon at the position indicated by Greek letters in Table IV, add the listed increment to 123.3. Add also the appropriate steric corrections. The allylic carbons of an alkene are calculated according to Table III for the saturated analog; then, if the alkene is cis, add −3, or trans, add +3. If both

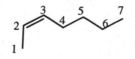), add 0. The remainder of the carbons are calculated by Table III for the saturated analog. *cis*-2-Heptene is predicted as follows:

| Increments | | Calculated | Measured |
|---|---|---|---|
| C-1: | 6.8 + 9.6 − 3.0 + 0.5 − 3 | = 10.9 | 12.6 |
| C-2: | 123.3 + 10.6 − 7.9 − 1.8 + 1.5 − 1.1 | = 124.6 | 123.5 |

| Increments | Calculated | Measured |
|---|---|---|
| C-3: 123.3 + 10.6 + 7.2 − 1.5 − 7.9 − 1.1 | = 130.6 | 130.8 |
| C-4: 15.3 + 2(9.8) − 2.7 + 0.3 − 3 | = 29.5 | 26.7 |
| C-5: 15.3 + 2(9.8) − 2.7 + 0.3 | = 32.5 | 32.0 |
| C-6: 15.3 + 9.8 − 2.7 + 0.3 | = 22.7 | 22.4 |
| C-7: 6.8 + 9.6 − 3.0 + 0.5 | = 13.9 | 13.9 |

2,4,4-Trimethyl-1-pentene is also predicted as follows:

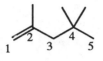

| Increments | Calculated | Measured |
|---|---|---|
| C-1: 123.3 + 2(−7.9) − 1.8 + 3(1.5) + 2.5 | = 112.7 | 114.4 |
| C-2: 123.3 + 2(10.6) + 7.2 + 3(−1.5) − 4.8 | = 142.4 | 143.7 |
| C-3: 15.3 + 16.7 + 21.4 | = 53.4 | 52.2 |
| C-4: 27.8 + 2.3 + 2(−0.9) | = 28.3 | 31.6 |
| C-5: 6.8 + 25.5 − 3.0 + 2(0.5) | = 30.3 | 30.4 |
| 2 CH₃: 6.8 + 17.8 − 3.0 + 3(0.5) | = 23.1 | 25.4 |

Replacement of a hydrogen with a heteroatom or a carbonyl group causes large shifts in the carbon to which it is attached and lesser changes

**TABLE IV. Parameters for Calculating $^{13}$C NMR Shifts of Vinyl Carbons**

$$\overset{\gamma}{C}-\overset{\beta}{C}-\overset{\alpha}{C}-C=C-\overset{\alpha'}{C}-\overset{\beta'}{C}-\overset{\gamma'}{C}$$

$\alpha + 10.6 \quad \beta + 7.2 \quad \gamma - 1.5 \quad \alpha' - 7.9 \quad \beta' - 1.8 \quad \gamma' + 1.5$

*Steric Corrections*

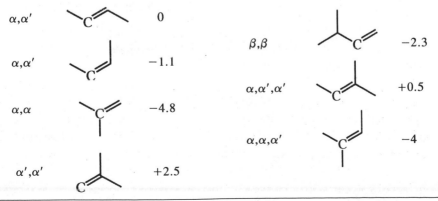

**TABLE V. Chemical-Shift Changes on Replacing a Hydrogen with the Group X[a]**

$$X- \overset{\overset{\displaystyle (C)}{|}}{\underset{\alpha}{C}} -\underset{\beta}{C}-\underset{\gamma}{C}$$

| X | $1°\alpha$ | $2°\alpha$ | $1°\beta$ | $2°\beta$ | $\gamma$ |
|---|---|---|---|---|---|
| —OH | 48 | 41 | 10 | 8 | −5 |
| —OR | 58 | 51 | 8 | 5 | −4 |
| —OCOR | 51 | 45 | 6 | 5 | −3 |
| —NH₂ | 29 | 24 | 11 | 10 | −5 |
| —F | 68 | 63 | 9 | 6 | −4 |
| —Cl | 31 | 32 | 11 | 10 | −4 |
| —Br | 20 | 25 | 11 | 10 | −3 |
| —1 | −6 | 4 | 11 | 12 | −1 |
| —COOH | 21 | 16 | 3 | 2 | −2 |
| —COOR | 20 | 17 | 3 | 2 | −2 |
| —COR | 30 | 24 | 1 | 1 | −2 |
| —CN | 4 | 1 | 3 | 3 | −3 |
| —Ph | 23 | 17 | 9 | 7 | −2 |
| —NO₂ | 63 | 57 | 4 | 4 | — |

[a]The 1° columns are used when the substituent is on a 1° carbon, and the 2° columns are used when the substituent is on a 2° carbon. Substituents on 3° carbons give irregular effects not amenable to simple empirical calculation.

to farther carbons. These effects are listed in Table V.[19] To use this table, first calculate the analogous hydrocarbon, using Table I, and then make the changes to the $\alpha$, $\beta$, and $\gamma$ carbons. For example, 3-methyl-1-butanol is assigned by first calculating the shifts of 2-methylbutane:

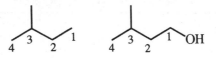

| | Increments | Calculated | Measured |
|---|---|---|---|
| C-1: | 6.8 + 9.6 + 2(−3.0) + 48 | = 58.4 | 60.7 |
| C-2: | 15.3 + 16.7 + 10 | = 42.0 | 41.7 |
| C-3: | 23.5 + 6.6 − 5 | = 25.1 | 24.8 |
| C-4: | 6.8 + 17.8 − 3.0 | = 21.6 | 22.6 |

In ketones, recognize that the whole COR is the substituent. For example, in calculating 3-heptanone the first two carbons are a substituted ethane, and carbons 4-7 are a substituted butane:

| Increments | | Calculated | Measured |
|---|---|---|---|
| C-1: | 6.8 + 1 | = 7.1 | 7.8 |
| C-2: | 6.8 + 30 | = 36.8 | 35.8 |
| C-3: | (Table VII) | | 211.2 |
| C-4: | 6.8 + 9.6 − 3.0 + 30 | = 43.4 | 42.1 |
| C-5: | 15.3 + 9.8 + 1 | = 26.1 | 26.2 |
| C-6: | 15.3 + 9.8 − 2 | = 23.1 | 22.5 |
| C-7: | 6.8 + 9.6 − 3.0 | = 13.4 | 13.8 |

The carbons of benzene rings may be assigned by starting with the value for benzene (128.5 ppm) and adding the parameters in Table VI.[19]

**TABLE VI. Chemical-Shift Changes at Each Ring Carbon on Replacement of a Hydrogen with Group X**

| X | 1 | 2 | 3 | 4 |
|---|---|---|---|---|
| —$CH_3$ | 9.3 | 0.8 | 0 | −2.9 |
| —$CH_2CH_3$ | 15.6 | −0.4 | 0 | −2.6 |
| —$CH(CH_3)_2$ | 20.1 | −2.0 | 0.0 | −2.5 |
| —$COOCH_3$ | 2.1 | 1.1 | 0.1 | 4.5 |
| —$COCH_3$ | 9.1 | 0.1 | 0 | 4.2 |
| —$OCH_3$ | 31.4 | −14.4 | 1.0 | −7.7 |
| —$OCOCH_3$ | 23.0 | −6.4 | 1.3 | −2.3 |
| —$NHCOCH_3$ | 11.1 | −9.9 | 0.2 | −5.6 |
| —$NH_2$ | 18.0 | −13.3 | 0.9 | −9.8 |
| —$NO_2$ | 20.0 | −4.8 | 0.9 | 5.8 |
| —Cl | 6.2 | 0.4 | 1.3 | −1.9 |
| —Br | −5.5 | 3.4 | 1.7 | −1.6 |
| —OH | 26.9 | −12.7 | 1.4 | −7.3 |
| —CN | −15.4 | 3.6 | 0.6 | 3.9 |
| —CHO | 8.2 | 1.2 | 0.6 | 5.8 |

As you might expect, electron-donating, ortho-, para-directing groups resonance-shield the ortho and para carbons, and the effects of any substituent are small at the meta position. Where more than one substituent is on the ring, the parameters are additive, as shown for 4-nitroanisole:

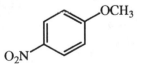

| Increments | | Calculated | Measured |
|---|---|---|---|
| C-1: | $128.5 + 31.4 + 5.8$ | $= 165.7$ | 164.7 |
| C-2: | $128.5 - 14.4 + 0.9$ | $= 115.0$ | 114.0 |
| C-3: | $128.5 + 1.0 - 4.8$ | $= 124.7$ | 125.7 |
| C-4: | $128.5 - 7.7 + 20.0$ | $= 140.8$ | 141.5 |
| $OCH_3$: | | | 55.9 |

In the proton-decoupled $^{13}C$ spectrum carbons 1 and 4 are apparent because they are quite small owing to incomplete relaxation and the lack of nuclear Overhauser effect.

The partial plus charge on carbonyl carbons gives them large chemical shift values as shown in Table VII. In conjugated carbonyl compounds the partial plus charge is delocalized and the signals are 8–13 ppm less than for nonconjugated carbonyl compounds.

More extensive tables are available, as well as other methods of calculating values.[19,22] A good collection of 500 $^{13}C$ NMR spectra is also available.[23]

Alicyclic ring carbons are not well predicted by the methods covered above, but they may be estimated by empirical correction methods based on analogous pairs of related molecules.[24]

## 10.4   CORRELATION OF $^1H$ AND $^{13}C$ NMR SPECTRA

A sequence of pulses at the proton and carbon frequencies can transfer part of the proton magnetization to the carbons with which they couple, along with the proton precession frequencies. Varying the delay between pulses gives a stack of spectra, which are Fourier-transformed to give a

**TABLE VII. Chemical-Shift Values (in ppm) of Common Carbonyl Carbons and Nitriles**

| | |
|---|---|
| Nitriles | 117–121 |
| Carbonates | 155–156 |
| Amides | 165–175 |
| Esters | 165–175 |
| Anhydrides | 165–175 |
| Acids | 173–185 |
| Aldehydes | 192–205 |
| Ketones | 197–220 |
| Acid chlorides | 168–174 |

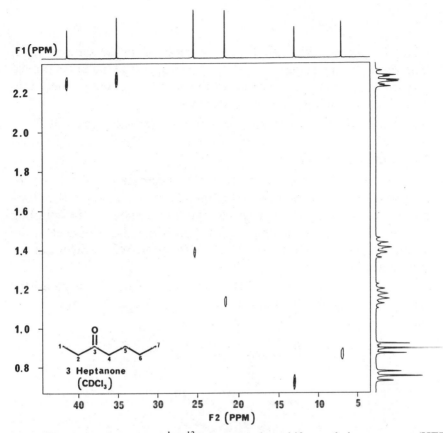

**Figure XV.** Two-dimensional $^1$H–$^{13}$C heteronuclear shift correlation spectrum (HET-COR) of 3-heptanone. Spectrum courtesy of Varian Associates.

two-dimensional spectrum (hetereonuclear coupling correlation, HET-COR) with the proton spectrum along one axis and the $^{13}$C spectrum along the other. Contoured peaks correlate the signal for a proton with the signal for the carbon to which it is attached. The HETCOR spectrum for 3-heptanone is shown in Fig. XV. Once again, the signals in 3-heptanone are assignable without HETCOR, but this simple example clearly shows the effect. As with COSY spectra, appropriate delays can be introduced in the sequence to give long range correlations. This can be optimized at about 140 Hz as in Fig. XV, or it can be optimized for 6 Hz for the longer-range couplings. These too contribute to the estab-lishment of the connectivity in a structure. A good example of signal assignments on the basis of long- and short-range HETCOR spectra can be seen in the work of Lankin and co-workers.[25]

## PROBLEMS

1. The $^{13}$C spectra in Fig. XIII were obtained from the same sample. The signal for the solvent $CDCl_3$ is the strongest in the undecoupled spectrum, but it is the weakest in the proton-decoupled spectrum. Why is there such a dramatic difference?

2. The $^{13}$C NMR of 4-methyl-2-pentanol shows six signals at 22.4, 23.1, 23.9, 24.8, 48.7, and 65.8 ppm. Why should we not expect five signals as in 2-methylpentane?

3. Figure XVI shows the 300-MHz $^1$H spectrum of a mixture of Z- and E-1,3,4,4-tetrachloro-1-butene. Draw the structures and assign every signal to the appropriate hydrogen. Show all the coupling constants in hertz, indicating which hydrogens they connect in the structures. Disregard the signal at 6.2106 ppm. What percent of the mixture is the Z isomer?

4. Figure XVII shows the 300 MHz $^1$H spectrum of **1**. Assign all the signals and explain the splitting patterns at 3.1 to 3.3, 5.0 to 5.1, and 7.51 ppm.

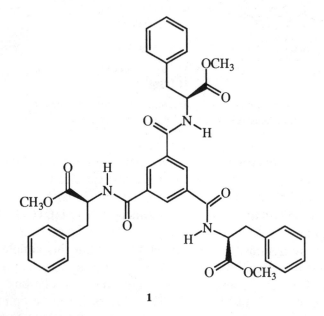

**1**

5. Examine the $^1$H NMR spectrum of acrolein diethyl acetal (Fig. XII), and assign all the signals to particular hydrogens in the structure. Calculate all the coupling constants, and indicate which hydrogens each one connects. Explain the coupling patterns.

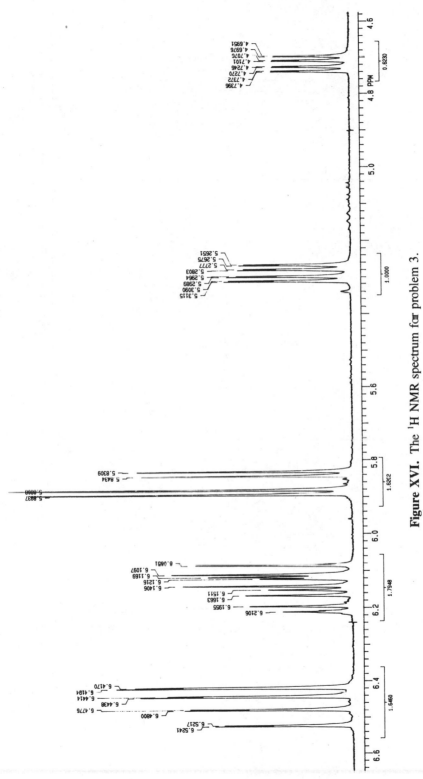

**Figure XVI.** The $^1$H NMR spectrum for problem 3.

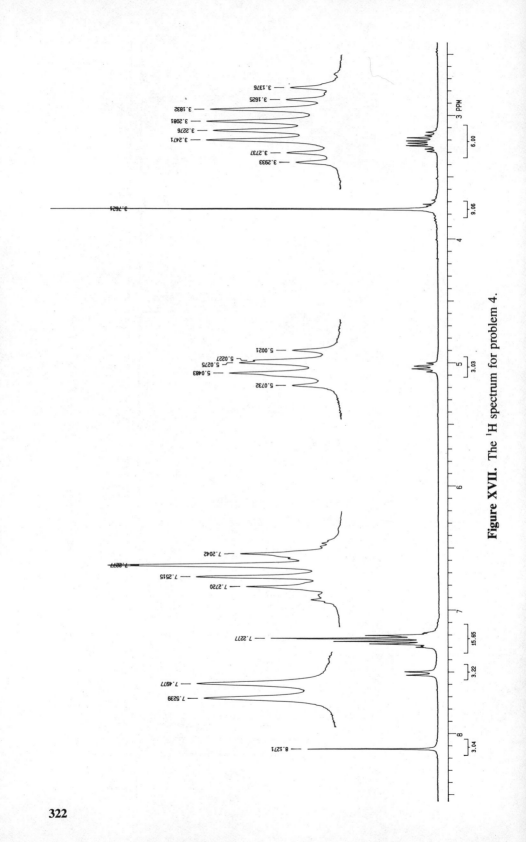

**Figure XVII.** The $^1$H spectrum for problem 4.

6. Treatment of 2,7-dimethyl-2,7-octanediol with pyridine hydrochloride gave many products, including three compounds with the boiling points and $^{13}C$ NMR spectra listed below.[26] Give the structure of each and assign as many signals as possible to particular carbon atoms in the structures.

   a. bp 162.5°C; 22.3, 27.3, 37.7, 109.6, 145.8 ppm

   b. bp 164°C; 17.7, 22.4, 25.7, 27.8, 28.0, 37.6, 109.7, 124.5, 131.1, 145.6 ppm

   c. bp 165.5°C; 17.7, 25.7, 28.6, 124.6, 131.1 ppm

7. Give the structure of the compound that gives the $^{13}C$ NMR spectrum in Fig. XVIII.

8. A certain compound of molecular formula $C_9H_{18}O_3$ dissolves in dilute sulfuric acid on prolonged boiling and gives off $CO_2$ gas. The proton-decoupled $^{13}C$ NMR spectrum of the compound is shown in Fig. XIX. What is the structure?

22.6 CH$_3$

24.4 CH

52.3 CH$_2$

210.0 C

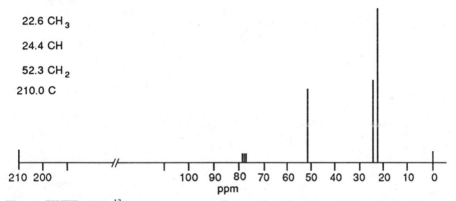

**Figure XVIII.** The $^{13}C$ NMR spectrum for problem 7. Adapted from L. F. Johnson and W. C. Jankowski, *Carbon-13 NMR Spectra*, copyright ©1972 by John Wiley & Sons, Inc. Reprinted by permission of John Wiley & Sons, Inc.

13.6

19.1

30.9

67.6

155.5 ppm

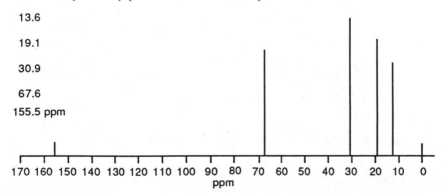

**Figure XIX.** The $^{13}C$ NMR spectrum for problem 8. Adapted from F. Johnson and W. C. Jankowski, *Carbon 13 NMR Spectra* L. copyright ©1972 by John Wiley & Sons, Inc. Reprinted by permission of John Wiley & Sons, Inc.

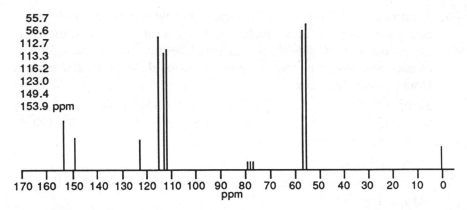

55.7
56.6
112.7
113.3
116.2
123.0
149.4
153.9 ppm

**Figure XX.** The $^{13}C$ NMR spectrum for problem 9. Adapted from L. F. Johnson and W. C. Jankowski, *Carbon-13 NMR Spectra*, copyright ©1972 by John Wiley & Sons, Inc. Reprinted by permission of John Wiley & Sons, Inc.

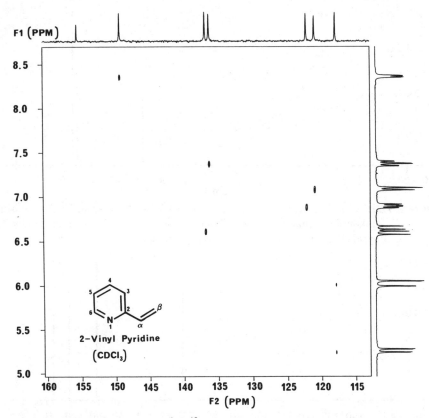

**Figure XXI.** Two-dimensional $^1H$–$^{13}C$ heteronuclear shift correlation (HETCOR) spectrum of 2-vinylpyridine for problem 10. Spectrum courtesy of Varian Associates.

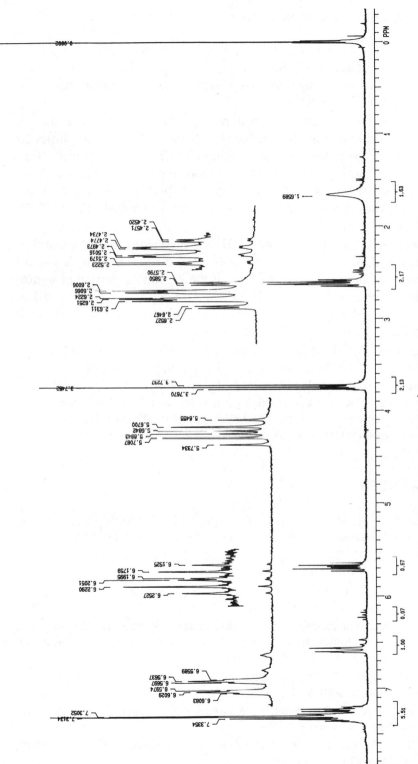

**Figure XXII.** The 300-MHz ¹H NMR spectrum for problem 11.

325

9. Fig. XX shows the proton-decoupled $^{13}C$ NMR spectrum of a compound of molecular formula $C_8H_9ClO_2$. Give the structure of the compound, and assign as many signals as possible to particular carbons in the structure.

10. Examine the $^1H$ splitting pattern and the COSY spectrum for 2-vinylpyridine (Fig. X) and assign all the $^1H$ signals to particular protons in the structure. Then, using the HETCOR spectrum (Fig. XXI) and the $^1H$ assignments, assign all the $^{13}C$ signals to carbons in the structure. Note that two different protons are on the same carbon, thus providing a sure starting point for analysis of the COSY spectrum.

11. Figure XXII is the 300-MHz $^1H$ NMR spectrum of 4-phenyl-3-buten-1-ol. It is mostly one geometrical isomer, along with a small amount of the other one. Assign all the signals to particular hydrogens in the structures, and show all the coupling constants in Hz, indicating which hydrogens they connect. Calculate the percent $Z$ and the percent $E$ isomers present.

12. The 90-MHz $^1H$ NMR spectrum of a certain compound showed an overlapping doublet and triplet, the individual maxima of which appeared at 1.254, 1.332, 1.357, 1.410, and 1.451 ppm. Calculate where these would occur on a ppm scale in a 300-MHz spectrum. Plot the appearance of the 90-MHz and 300-MHz spectra on scales of 1 cm = 0.010 ppm.

## REFERENCES

1. Silverstein, R. M.; Bassler, G. C.; Morril, T. C. *Spectrometric Identification of Organic Compounds*, 4th ed., Wiley, New York, 1981.

2. Cooper, J. W. *Spectroscopic Techniques for Organic Chemists*, Wiley-Interscience, New York, 1980.

3. Williams, D. H.; Fleming, I. *Spectroscopic Methods in Organic Chemistry*, 3rd ed., McGraw-Hill Book Company (UK) Ltd., London, 1980.

4. For an excellent example of the value of laser Raman spectroscopy, see Smith, L. M.; Smith, R. G.; Loehr, T. M.; Daves, G. D., Jr.; Daterman, G. E.; Wohleb, R. H. *J. Org. Chem.* **1978,** *43*, 2361.

5. McLafferty, F. W. *Interpretation of Mass Spectra*, 3rd ed., University Science Books, Mill Valley, CA, 1980.

6. Bovey, F. A. *Nuclear Magnetic Resonance Spectroscopy*, 2nd ed., Academic Press, New York, 1988.

7. Williams, K. R.; King, R. W. *J. Chem. Ed.* **1989,** *66*, A213, A243; **1990,** *67*, A125, A213.

8. Farrar, T. C.; Becker, E. D. *Pulse and Fourrier Transformation NMR*, Academic Press, New York, 1971.

9. Macomber, R. S. *J. Chem. Ed.* **1985**, *62*, 213.

10. Derome, A. *Modern NMR Techniques for Chemistry Research*, Pergamon Press, Oxford, 1987.

11. Corio, P. L. *Structure of High-Resolution NMR Spectra*, Academic Press, New York, 1966.

12. Emsley, J. W.; Feeney, J.; Sutcliffe, L. H. *High Resolution Nuclear Magnetic Resonance*, Pergamon Press, New York, 1965.

13. Grunewald, G. L.; Ye, Q. *J. Org. Chem.* **1988**, *53*, 4021.

14. Brey, W. S., Ed., *Pulse Methods in 1D and 2D Liquid-Phase NMR*, Academic Press, New York, 1988; Schraml, J.; Bellama, J. M., *Two-Dimensional NMR Spectroscopy*, Wiley-Interscience, New York, 1988; Morris, G. A., *Magn. Reson. Chem.* **1986**, *24*, 371–403.

15. Cho, B. P.; Harvey, R. G. *J. Org. Chem.* **1987**, *52*, 5679.

16. Portlock, D. E.; Lubey, G. S.; Borah, B. *J. Org. Chem.* **1989**, *54*, 2330.

17. Cushman, M.; Chinnasamy, P.; Patrick, D. A.; McKenzie, A. T.; Toma, P. H. *J. Org. Chem.* **1990**, *55*, 5995.

18. Fraser, R. R.; Petit, M. A.; Miskow, M. *J. Am. Chem. Soc.* **1972**, *94*, 3253.

19. Wehrli, F. W.; Marchand, A.; Wehrli, S. *Interpretation of Carbon-13-NMR Spectra*, 2nd ed., Wiley-Interscience, New York, 1988.

20. Pegg, D. T.; Doddrell, D. M.; Bendall, M. R. *J. Chem. Phys.* **1982**, *77*, 2745.

21. Lindeman, L. P.; Adams, J. Q. *Anal. Chem.* **1971**, *43*, 1245.

22. Brown, D. W. *J. Chem. Ed.* **1985**, *62*, 209.

23. Johnson, L. F.; Jankowski, W. C. *Carbon-13 NMR Spectra*, Wiley-Interscience, New York, 1972.

24. Petiaud, R.; Taarit, Y. B. *J. Chem. Soc. Perkin II* **1980**, 1385.

25. Shieh, H.-L.; Cordell, G. A.; Lankin, D. C.; Lotter, H. *J. Org. Chem.* **1990**, *55*, 5139.

26. Dauzonne, D.; Platzer, N.; Demerseman, P.; Lang, C.; Royer, R. *Bull. Soc. Chim. Fr. II*, **1979**, 506.

# INDEX